COMPUTER-BASED CHEMICAL INFORMATION

BOOKS IN
LIBRARY AND INFORMATION SCIENCE

A Series of Monographs and Textbooks

EDITOR
ALLEN KENT

Director, Office of Communications Programs
University of Pittsburgh
Pittsburgh, Pennsylvania

COMPUTER-BASED CHEMICAL INFORMATION

Edited by

Edward McC. Arnett

Department of Chemistry
University of Pittsburgh
Pittsburgh, Pennsylvania

and

Allen Kent

Office of Communications Programs
University of Pittsburgh
Pittsburgh, Pennsylvania

MARCEL DEKKER, INC., New York   1973

MARCEL DEKKER, INC.
95 Madision Avenue, New York, New York 10016

LIBRARY OF CONGRESS CATALOG CARD NUMBER: 72-95840

ISBN: 0-8247-6045-X

Printed in the United States of America

PREFACE

This book is a report of a large, multidisciplinary study on the implementation and use of computer-based chemical information, completed in 1970. In addition to the authors of the various chapters, many other colleagues were involved in the project. We would like to express our particular appreciation to the following people: Mrs. Mary Jane Bloemeke, Chemistry Libriarian, without whom the initial plan for the project would not have been developed; Mary Jane Pugh and Betty Hartner who helped at every stage in the development of workshops and profiles; Dr. Elaine Caruso for the development and testing of interactive tutorials for self-instruction in profile development; Dr. Siegfried Treu for much patient assistance in systems development; James Weir, Vince Sefcik, Jan Jones, and Griffith Smith for application programming; Edmond Howie for development and maintenance of liaison with industrial users; Dr. Carl Beck and Dr. Jerome Laulicht for guidance of the behavior research effort; and Mrs. Donna Williams for steady administrative support during the entire operation of the project. Finally we acknowledge the support of the National Science Foundation (grant GN738), the Pennsylvania Science and Engineering Foundation, and the University of Pittsburgh, without which this study would have been impossible.

Edward McC. Arnett
Allen Kent

LIST OF CONTRIBUTORS

DANIEL JAMES AMICK, * Director, Behavioral Task Group, Pittsburgh Chemical Information Center, University of Pittsburgh

EDWARD McC. ARNETT, Professor of Chemistry, University of Pittsburgh

ELAINE CARUSO, Director, Interactive Applications Task Group, Pittsburgh Chemical Information Center, University of Pittsburgh; Assistant Professor, Interdisciplinary Doctoral Program in Information Science, University of Pittsburgh

BAHAA EL-HADIDY, Chemical Information Specialist, Pittsburgh Chemical Information Center, University of Pittsburgh; Knowledge Availability Systems Center, University of Pittsburgh

NEALE S. GRUNSTRA, Project Manager, Pittsburgh Chemical Information Center, University of Pittsburgh; Director, Administrative Systems, University of Pittsburgh

K. JEFFREY JOHNSON, Assistant Professor of Chemistry, University of Pittsburgh

ALLEN KENT, Professor of Library and Information Sciences, University of Pittsburgh; Director, Office of Communications Programs, University of Pittsburgh

*Present address: Department of Sociology, University of Illinois at Chicago Circle, Chicago, Illinois

v

CONTENTS

Chapter 3   THE USER'S INFORMATION SYSTEM: AN
             EVALUATIVE RESEARCH APPROACH   43

Daniel James Amick

Chapter 4   SYSTEM DESIGN, IMPLEMENTATION, AND
             EVALUATION   83

Neale S. Grunstra and K. Jeffrey Johnson

Chapter 5   INTERACTIVE RETRIEVAL SYSTEMS · · · · · · · ·   125

Elaine Caruso

COMPUTER-BASED CHEMICAL INFORMATION

Chapter 1

THE RESEARCH CHEMIST AND HIS
INFORMATION ENVIRONMENT

Edward McC. Arnett and Allen Kent

Department of Chemistry
University of Pittsburgh
Pittsburgh, Pennsylvania

and

Graduate School of Library and Information Sciences
Office of Communications Programs
University of Pittsburgh
Pittsburgh, Pennsylvania

1

I. INTRODUCTION

From one point of view, the activities of the research chemist are totally concerned with chemical information—its procurement, evaluation, and dissemination. This point of view has been fostered by the chemist historically—by actions if not by words. Long before the professional societies in most other fields had paid significant attention to the information function, those in chemistry were providing mechanisms for original publication of research results, for repackaging of this information in the form of abstracts, and for providing ready access to this information through the development of indexing services.

Therefore, the information scientist who has undertaken the task of developing more effective and efficient tools and services for accessing information, with the aid of computers, has sometimes been puzzled and disappointed when his efforts have been met with apathy or hostility from those whom he wishes to serve.

It, therefore, seemed appropriate for a research chemist to intrude into the world of information science, and to learn more about the interface between chemistry and information services. This was especially timely when more and more resources were being allocated to the processing of information for chemists, seemingly at the expense of support for the research that produced the original information.

The Chemical Information Center Experiment Station was established specifically for the implementation of computer-based chemical information services in Pittsburgh and to investigate their acceptance by a broad and representative range of research chemists. The team assembled to lead this activity included a research chemist, an information scientist, a librarian, a behavioral scientist, and a computer scientist.

Detailed descriptions of a number of aspects of our experience are described in other chapters in this book. However, it seems important to describe first, as candidly as possible, the initial attitudes of the research chemist, who served as principal investigator, and of the information scientist, who helped to bring the necessary resources to the project. These attitudes relate to areas of mutual interest and of conflict.

The immediate perspective of the information scientist is taken from the vantage point of an academic research person turned administrator who several decades ago was a chemist. His conversion to information science was both dramatic and evolutionary. It was dramatic in terms of the real-time change from working with chemicals to working with information. It was evolutionary in terms of the piecemeal unfolding of a gradual understanding of the factors that need to be taken into account in the field.

Information science did not emerge as a separate discipline until very recently. In the beginning, there was the librarian who was responsive to the need to acquire and store recorded knowledge, and to keep track of materials missing from the library shelves. Then, there was the documentalist, who began to grasp the fact that a more dynamic role could be taken in understanding the needs of the library clientele and in serving them. This was perhaps the start of active dissemination of recorded information, sometimes in anticipation of a need as well as in response to a request. The next step was the development of the science information specialist (in chemistry often called the "literature chemist"). This specialist—part scientist, part librarian or documentalist—assumed various roles. In some cases, he was a special librarian with training in chemistry; in other cases he assumed the literature-searching function of the chemist. A few specialists became part of an elite group who were expected to be steeped in the literature of chemistry and to participate in research planning with counterparts who specialized in experimental work. The next evolutionary step was the emergence of the information scientist who, as the name implies, studies information as a scientific phenomenon in its own right.

The immediate perspective of the principal investigator cannot help being personal, but is taken from the vantage point of an academic research chemist with some industrial background, who has been immersed for the better part of four years in chemical information activities. This involvement came about through an intense feeling of need for assistance in accessing and organizing chemical literature systems. He believed that those versed in the use of computers for organizing and handling large masses of information should be able to help the research chemist, and considered that as a teacher he needed to prepare himself and his students for the services which would probably be available during their professional lifetimes. Perhaps the testimony drawn from one man's experience would be worth something, but it alone should not justify a chapter in a book of this kind at this time. Hopefully, the reactions described in this chapter are a balanced report based on many discussions with chemists at the University of Pittsburgh, Mellon Institute, Harvard University, and in a number of universities in the United Kingdom.

A. Research Chemistry and Its Aims

Chemistry is that area of the natural sciences which deals with the structure and interconversion of materials. Therefore, the relationship between matter and energy is always paramount in chemistry. Some chemists, such as X-ray spectroscopists, dedicate their lives purely to

the determination of molecular structure and others in thermodynamics
or kinetics are purely concerned with determining relative energies. However, these activities are always carried out within the broader context of
chemistry, which is always concerned with the interconvertibility of matter.

There is much discussion these days about differences between basic
and applied research. We consider that applied research is carried out
deliberately towards the solution of some practical problem or for the
development of a process or new material. In contrast, basic research is
generally directed towards more open-ended questions, either in the ex-
ploration of an insufficiently understood relationship between different
types of matter or energy, or the testing of some fundamental principle
involving such relationships.

In view of the millions of dollars which have gone into the development
of chemical information systems, we have considered that it is of great
importance for planning future computer-based services to identify as best
we can those scientists who are most apt to be aided by these systems and
to concentrate our efforts in assisting them.

In considering different ways of regarding research and the research
styles of scientists, perhaps the most useful way lies along a continuum
from applied research to basic research. And although many classifica-
tions and ways of looking at research scientists have not stood up to careful scrutiny, it does appear quite certain that industrial applied researchers
are served more readily by computer-based information than are many
types of chemists in basic research. The important reasons for this are
fairly obvious and have been recognized for a long time, as is clear from
the fact that many of the major companies employ large retinues of science
information specialists and/or information scientists and are perhaps a
generation ahead of academia in the use and acceptance of computer-based
information service. By contrast, the majority of their academic colleagues are almost completely innocent (or ignorant) of the existence of
these services, let alone the fine points of their use. Although some
American academicians would argue that the recent funding crisis for
their research is primarily to blame for this situation, a similar lag between industrial and academic researchers in the United Kingdom is also
found, even though the academic community of the United Kingdom has been
carefully prepared for the introduction of current-awareness services
through a well-planned campaign over the last several years.

Industrial organizations and the research which is carried out in them
generally has clearly defined goals and the clarity of that definition is
directly related to how applied the project may be. It may also be related
to the size of the company in terms of number of employees, gross sales,
and profits. There are in the United States a relatively small number of

companies large enough to support research institutes where basic research in areas related to company goals is conducted. Such research not only leads to new products and new areas for the long-range growth of the company but also provides a cadre of internal experts who can serve as consultants to research groups within the company that are working on more applied projects. This not only serves to attract first-class scientific brains to the company, but allows it to move rapidly in new directions without consulting outside experts. The public view of industrial research is very frequently colored more than it should be through the public relations image projected from high-prestige fundamental groups within a small number of large and well-to-do industrial concerns.

Although a few large companies can support such basic research groups, most of the research carried out in the industrial community (including most of the research done in large prestigious companies) is directed toward sharply defined goals within areas that can be quite narrowly described and are therefore readily classified for the storage and retrieval of the information concerning the area. For example, a company which does 90% of its business in phenol formaldehyde resins for electrical components, should be able to describe very accurately the type of library which it should have and the types of computer-based services which will handle a very large portion of its needs. At the other end of the basic-to-applied spectrum, one may find in industry or academia, the theoretical chemist who considers himself to be an expert in quantum mechanical or statistical mechanical calculations and whose broad information needs may be very difficult to define.

B. Some Varieties of Research Style

The use to which research chemists put computer- processed information is, in part, determined by the goals of their employers (industrial versus academic institutions) and by the type of field in which they find themselves—such, for example, as chemical engineering in contrast to quantum mechanics. However, we have found that the degree to which an individual research scientist is attracted to the use of computer-based services and the satisfaction which he gets from them depends in many cases on behavioral factors. These are much harder to describe than the simple applied-versus-basic or academic-versus-industrial dichotomies. Scientists and those who deal with scientists have a clear impression that there is a wide variety of personality or character types within the scientific community. This is popularly described in terms of a degree of eccentricity that is often a source of amusement or irritation to those attempting to serve the researcher. This intuitive impression that there is a wide variety of scientific types is completely supported by the results of our

behavioral group and shows up as an enormous variance in the responses to
the behavioral instruments which have been used in studying our user
groups.

This variance implies that anyone will be making a grievous error if
he deals with the scientific community on a monolithic basis by forcing
creative individuals to conform to systems that have been developed primarily for convenience in machine processing. Either he will be rejected
by the scientists through constructing an unaccepted system with a conse-
quent loss of time and money; or, if he attempts to force them to use a
system they do not want, he may do irreparable damage to the very creative forces he is attempting to assist.

Perhaps the personality component most conducive to the development
and acceptance of information storage and retrieval systems for the research scientist is the degree to which he is systematic and maintains a
long-range structural interest in a given field. The scientist whose life
work is devoted to the study of infrared spectra of heterocyclic compounds
is almost forced to maintain a good filing system in which the structures of
the molecules are related systematically to their infrared spectra. Such a
man, by the nature of his field, must be highly systematic, but this does
not necessarily mean that he is attracted to or will accept a system developed by an outside expert, such as an information scientist. Indeed, he
may be so pleased with his own system that he feels it could scarcely be
improved on by an outsider and will not wish to spend the time or money
in changing it. From a less positive point of view, he may also resent and
feel threatened by the implication that he may not have been doing his job
as perfectly as it could be done and will not want to hear about what he has
been missing.

On the other hand, another highly organized scientist who maintains a
well-structured relationship to his field and who appreciates the development of systems may realize clearly that much of the time he spends in
organizing his files could be saved by the use of computer-based systems.
He will seek out those who can supply such systems to him. If he makes
this effort, and particularly if he takes the initiative in paying even a
minimum price to obtain such services, evidence indicates that he will
probably obtain satisfaction from them, make the effort to keep them up,
and will be ready to pay a fair portion of his research budget to continue
getting such service. This type of scientist will profit well from current-
awareness services, since he knows his field and is able to develop the
necessary interest profiles that can be used for accessing presently available services through key words, authors, or citations. Such a man's
private files will probably be well maintained and readily available for the
development of systems for information storage and retrieval. Interactive accessing of commercially available infrared files may be of value to

such a man and he may wish to develop an interactive file based on the spectra from his own research.

Another type of scientist may be classified as the "hit-and run" character type whose research problems arise on a rather ad hoc basis, coming from a very general area of interest such as synthetic-organic chemistry. Such a person's research may be directed towards clearly articulated goals, such as the synthesis of particular compounds. But once a successful synthesis has been achieved, he will move on to something else, maintaining little, if any, interest in the area in which he was previously completely involved.

The maintenance of long-term files or current-awareness profiles has little attraction for researchers whose style is purely of this type. Nevertheless, he will certainly need to be inspired by browsing through journals which have proved valuable in the past. He may need to be alerted regularly to what the leaders in his field are doing and perhaps maintain a structure profile on particular compounds he is trying to prepare. If he is moving rapidly into a new field, a retrospective search may be of great importance to him. Since the wide use of computer-based retrospective searching is only in its infancy, we have not had an adequate opportunity to document this impression, but by the nature of the case it would seem to be true.

A person's research style may vary from time to time. Also, several different styles can be discerned in a single individual who may have one or two long-range interests which he will maintain in a systematic way while at the same time making a variety of hit-and-run forays into new fields which interest him.

Most active research scientists enjoy innovation and exposure to new ideas, so that even if they were left entirely to themselves in a static field the demands of their intellectual growth and curiosity would involve them in fields or techniques which were new to them. Accordingly, their interest profiles would be continually changing. Furthermore, the interest profile of an active scientist will be also changing continuously due to the rapid growth of the fields which interest him. These will be generating new vocabularies and the structures of new compounds as a natural part of their development. Accordingly, a scientist's interest profile must be revised constantly as his personal interests shift in various directions, as new authors enter the field of his interest, and as new words are invented in the vocabulary of the field.

In addition to such changes in the scientist's information needs and acceptance of them, it is important to realize that a research scientist, like any other person, responds to information stimuli that are given to him in accordance to his mood at the time when he is evaluating the information presented to him. If he is tired, harassed, and there is much competition for his attention from large stacks of mail, unread literature, or administrative tasks to be done, his attitude to being alerted to new literature may be hostile, and a journal or printout of current-awareness alerts which might be studied with care and enthusiasm on one day, could be dealt with in a very cursory manner on another.

II. THE AIMS OF CHEMICAL INFORMATION SCIENCE

Information science is concerned with the nature, organization, and transfer of information. As the term is applied to chemical information, the emphasis is primarily on the organization, storage, retrieval, and transfer of information rather than upon its acquisition. This is the case, because the main burden of acquiring published information is carried by the relevant professional societies and commercial information services. Thus, the activities of the chemist in the laboratory are aimed at the development of strategies and instruments for gathering information from natural systems, but it is in the library or the computer center (rather than the laboratory) that the chemical information expert is primarily employed. His concern, therefore, is not directly with the behavior of molecules or of natural systems, but with the manipulation of symbols which the scientist has already decided contain information about chemical systems and molecules. The chemical information scientist works with words or numbers, and it is for this reason that his role is often regarded by the research chemist as a secondary one.

However, the chemical information specialist, doing the bidding of the research chemist in conducting information searches based on his prescription, might, perforce, become interested in molecular structure and the theory of coding, and eventually develop advanced notation systems that can serve the chemist more effectively than he can serve himself.

In addition, the information specialist may wish to investigate how research might be conducted by manipulation of information per se. He may ask himself, for example, whether it is possible to predict the outcome of a synthesis not yet performed from the evidence derived from the synthesis of homologous compounds. The predictions, of course, could not be permitted to stand without the review of the creative chemist, but perhaps likely areas for attention might thus be uncovered.

III. COMPONENTS OF THE CHEMICAL INFORMATION ENVIRONMENT

Chemical information is stored in a variety of forms; for example, spectra, books, tabulations, magnetic tape, and knowledge in people's minds. Some types of information, such as the indexes derived from the titles of articles or books, are relatively amenable to processing in a digital computer and thus can be handled easily for transfer from one person to another through the agency of the information specialist. Other types of information may be so abstract and require so much training to understand that they can only be communicated directly from one expert to another through face-to-face discussions. This type of communication within the "invisible college" at seminars, conferences, meetings, and classroom discussions is necessary for the development and continued growth of all research scientists. It may suffice as a source of information for a few.

The value of the invisible college as a means of information transfer depends on at least three primary factors. Of these, probably the most important is the sophistication of the participants in the discussion. Thus, within a scientific elite (such as the chemistry section of the National Academy of Sciences or within the community of Nobel Laureates), it is likely that important information will be transmitted and received efficiently. Members will have a common recognition of what is important in their areas, and will intuitively gauge the level at which conversation should be carried out from what they know of the field, the other man, and the way he responds. Such information transfer clearly depends upon the degree to which the transmitter and the receiver are mutually attuned.

The other two factors which have important bearing on the value of the invisible college for information transfer have been alluded to already. They are a) the field of interest, and b) the character and research style of the scientists involved. In our experience, it appears that scientists engaged in theoretical, highly speculative, and unstructured research which is much more dependent upon concepts than upon accumulated data, will probably profit most from communication inside the invisible college. Such scientists may have attitudes of mistrust or even contempt for the published literature. The information scientist may do well to adopt an attitude of live and let live towards scientists of this type, since there is probably little he can do for them professionally. Correspondingly, it is important that men and women who are not in areas where data storage and retrieval are of much importance should not be much involved in making decisions regarding facilities and systems which may be of great value to chemists in other fields or of other research types.

Of the various components in the chemical information environment, none appears to be of greater importance to the research scientist than is the chemical journal. This is the permanent record of knowledge in his field, to which he must turn in order to develop any sort of retrospective search for concepts, data, or information on how to carry through a synthesis. Thus, a good chemistry library is evaluated very much in terms of the completeness of the back runs of its journals, as well as its current subscription list. Again, the importance of journals, and particularly back journals, will vary considerably depending upon the field and the character type of the chemist involved. Obviously, nuclear chemists have relatively little dependence upon the early chemical literature, whereas synthetic chemists may find valuable information in literature going far back into the nineteenth century.

It is from the journals that the best knowledge of advances in research can be found. Through the leading review journals, such as Chemical Reviews, Accounts of Chemical Research, and Quarterly Reviews, the scientist can keep abreast of fields peripheral to his own and can initiate an effective retrospective search when approaching a new field. The outstanding journals of general interest, such as the Journal of the American Chemical Society, the Journal of the Chemical Society of London, and Chemische Berichte, provide the environment for a degree of browsing, by means of which the chemist may keep abreast of the leading work in his field and may see the "hottest" new research as it appears. Articles in first-rate chemical journals are reviewed carefully by referees, and the rejection rate is high enough to insure the quality of most of the papers that are published. This provides a degree of reliability for the permanent record, but also means that papers published in leading journals may describe work submitted for publication nine months to a year beforehand, and which may have been completed in the laboratory two or three years before that. For the majority of chemists, however, the primary journal is the principal point of access to the overall field of chemistry, and is irreplaceable.

The scientist who has earned a recognized position in an area of his field will have, in addition to the face-to-face discussion within the invisible college, a variety of other points of contact to work going on in his field. Thus, he will serve as referee for articles submitted to the journals, and thereby have access to information which is only available to the general public six or eight months later. He will referee proposals for research grants or contracts which are submitted to agencies of the Federal government, such as the National Science Foundation or the National Institutes of Health. If he is active in one of the new and self-conscious areas of emerging science composed of a small community of recognized experts who know each other, he may participate in the exchange of preprints or newsletters which will never become part of the public record.

IV. INFORMATION SERVICES FOR THE RESEARCH CHEMIST

It is, or should be, the aim of the chemical information scientist to aid research chemists in the storage, retrieval, and transfer of the various types of chemical information which are amenable to systematic manipulation. Traditionally, this has been handled through the chemistry library and the various hard-copy indexing services, such as Chemical Abstract

Services and Chemisches Zeitschrifte. Through these, most scientists carry out their own literature searches and depend for part of their current-awareness needs on regular visits to the library to examine recent additions to Chemical Abstracts. Most researchers have been left to their own devices to develop systems for storing bibliographic citations, abstracts, and reprints of their important papers, as well as for tabulating the data which accrue from year to year in their own research.

With the constant growth of the chemical literature, it is not surprising that many conscientious scientists are appalled at the prospect of carrying out a complete retrospective search through the evergrowing indexes of Chemical Abstracts and Science Citation Index. They are also worried about their ability to keep abreast of their field, through watching the ten or fifteen journals of greatest interest to them, and wonder how much they are missing which could only be caught by constant surveillance of Chemical Abstracts or some other current-awareness periodical that reports on journals peripheral to their interests. They realize that to do a thorough job of "keeping up with the literature" would completely consume a forty-hour week. At the end of that time they would have accomplished nothing in the laboratory and would still be behind the flood of new literature published during that week.

In addition to worrying about externally-generated information in the published literature, the scientist will be concerned about the rapid growth of his own files. To protect himself, he will devise more intricate means for cross-indexing his reprints so that he can retrieve the papers relevant to a given article when the time comes to write up his research in a given field. If he is conscientious and methodical, he will probably find himself surrounded by ever-increasing rows of notebooks and files which are impossible to cross-index completely; and he will probably have made one or more unsuccessful attempts to improve the cross-indexing by assorted punch-card systems. Within this context, he may well feel that his time is becoming so occupied with housekeeping and bookkeeping that little time is left for the consideration of scientific problems. In this case, he may withdraw to a different field, settle for doing an inadequate job, or turn (directly or indirectly) to information scientists for help.

The computer-based current-awareness systems of Chemical Abstract Services, the Institute for Scientific Information, and a variety of other sources can now supply a very complete and virtually foolproof current-awareness system for the scientist who is ready to work with an information specialist at constructing a suitable interest profile. Although these systems are currently too expensive to use routinely for retrospective searches in the development of automatic bibliographies, research is being carried out on providing this service; and it seems likely that in the near future computer-based retrospective searches through Chemical Titles or

CA Condensates tapes may be applied with considerable value to different stages in the development of a complete bibliography for a man or woman entering a new field.

The development and use of massive data bases, such as Chemical Titles, ISI, or CASCON, depends upon participation by a large number of users in order to supply the necessary financial support. The other problem of cross-indexing, storing, and retrieving the information in a scientist's own data and bibliographic files has also been approached by the information scientist and some experiments in the Pittsburgh Chemical Information Center.

V. THE INFORMATION SCIENTIST, FRIEND OR FOE ?

Having considered the activities of the research chemist and of the chemical information scientist, we may now turn to an examination of those areas where they may work with mutual benefit, and, perhaps more importantly, consider where they may be in conflict. This may be approached most profitably if we assume that from the viewpoint of most research chemists, the only thing that the chemical information scientist has to offer is greater efficiency in dealing with the chemical literature or with the chemist's own files, thereby saving him money, time, or both. The chemist who wishes to use information systems has no interest in how those systems work and will wish to have as little to do with them, and with information scientists, as he possibly can. In this sense, his attitude will be very similar to that which he has in using a complicated scientific instrument, which he wishes to approach as a black box. His only concern is that he will receive the answer which is asked of the system or instrument as quickly and as cheaply as possible. He will have as little interest for the details of programming, coding, or Boolean logic, as for the detailed circuits of his spectrophotometer.

For his part, the chemical information expert may enter his field from a variety of areas—through library science, computer science, linguistics, or systems analysis. If he has had chemical training, or perhaps even been a chemist in the past, his commitment to the chemical information field is apt to be so completely absorbing that he no longer can keep abreast of chemistry, or think as a research chemist does. If he is a true professional, his chosen area is just as exciting to him and just as interesting as chemistry is to the research chemist. But he will be making a serious error if he thinks that he can interest the chemist in the details of what he is doing. On the other hand, just as the chemist cannot solve all of the riddles of chemistry, it is also true that the information scientist cannot solve all of the problems of information. All the table-pounding in the world cannot force the chemist to synthesize acetaldehyde from methane and sulfuric acid. Similarly, good information cannot be developed from poor data, and good information services cannot emerge without effective interaction between the system and the chemist. Effective interaction cannot be produced without a penetrating study of the behavior of chemists; furthermore, if information transfer is to be accomplished, the possibility must be kept alive that the chemist may change his behavior in relation to both the use of information and the use of new systems in novel ways, which may at first appear threatening to him. The introduction of the mass spectrometer changed the behavior of the analytical chemist by opening new vistas in his field. Hopefully, information systems may also provide such opportunities.

In the previous section, a number of areas have been identified in which modern information science, using computers to manipulate chemical information, can be of value to the research chemist. What acceptance computer-based chemical information has enjoyed from the scientific community certainly grows from common interest in what the information specialist can do for the scientist. However, it is clear to anyone attending meetings of information experts that their acceptance is by no means universal, and that they are frequently puzzled by their rejection from the scientific community.

The information scientist may be considered a threat by the members of the scientific community he wishes to serve if they think that he wastes their time or spends money that could be available to them. Competition for funds may occur at the local level where within a department a decision must be made as to whether to buy a new mass spectrometer, to improve the chemistry library, or to purchase information profiles or retrospective searches. In industrial organizations, the competition for such funds may not take place within the departmental research budget. The library is supported independently, so that searches, both retrospective and current-awareness, may be purchased without obvious cost to the research chemist. If he is suddenly faced, on the other hand, with the prospect of dipping into his own research budget to subscribe to a current-awareness service, he may tell himself that he got along quite well for many years without it, and that he will continue to do so. This will probably be true even if he is accustomed to using part of his research budget (or even personal resources) for journal subscriptions or books. At another level, the competition for funds may be manifested in hostility towards the activities of granting agencies which use some of their money for the support of information systems at the same time that the support of fundamental research is dwindling. It is not surprising if the chemist who has just lost two-thirds of his research budget through the failure of the National Institutes of Health to renew a grant in physical organic chemistry takes a negative attitude towards the development of an expensive new system for structure-searching.

Of equal, or perhaps greater, importance than the expenditure of money is the wasting of time. This must include infringements on what is perhaps the most crucial element to any creative individual, the distraction of his attention. The proper use of a computer-based information service requires a considerable investment of time—for the preparation and revision of profiles, discussion of the problem with an information expert, and evaluation of computer printouts which generally contain a major percentage of irrelevant material and frequently miss important papers. Once the profile is working properly, a variety of other irritations appear. Thus, there are a number of waiting periods, which are frequently quite inconvenient for the scientist being served. If he uses an "early warning system," such as ASCA or <u>Chemical Titles</u>, he may be informed of papers long before the journals carrying them arrive in his library. At the other extreme, using <u>CA Condensates</u>, he may be alerted to an important paper six or seven months after it has been published, due to the time lag inherent in the abstracting system and publication. He thus loses the feeling of first-hand contact with the literature which he had previously obtained from flipping through the pages of his favorite journals himself.

If the development of a literature system is carried to an idealized extreme, through the proper development of a good system of library support, the scientist will be alerted to everything that might be of interest to him. He will receive a collection of just those papers which may be of value to him at the time they arrive at the library. This will include alerts to papers published in journals to which his library does not subscribe. As far as the library scientist is concerned, he has done his job and done it well. However, the recipient of the information may discover, to his dismay, that his situation has changed from being under-informed to being over-informed. He is now faced with the problem of assimilating the wealth of information that has been provided to him. He may feel obliged to follow up a lot of alerts. Frequently these alerts are located in obscure journals, which he will obtain, sometimes at considerable cost, through inter-library loans, only to find that they are of little value to him.

The chemical information specialist, like the research chemist, believes in what he is doing, and may inflate the importance of it, especially when addressing his colleagues. Thus, one frequently hears from information scientists of research chemists who had depended upon the invisible college and "the grapevine" for keeping abreast of their field and who, after entering a computer-based information system, are embarrassed to learn that they had missed many references related to their field. However, the information scientist should realize that although many of these articles are "technically" relevant to the chemists' interests, many are not relevant from a practical point of view, and the chemist does not wish to spend time reading them.

This brings the information scientist to one of the most challenging research areas of his discipline: the nature of relevance judgments. It again suggests the importance of studying the process of information transfer in real-life situations—so that the scientist whose behavior is under scrutiny is studied "in vivo." Thus, every effort should be made to see that relevance judgments are based on real value to the user rather than on what the information expert considers to be "technically relevant."

In supplying scientific information the important question often may be, how much is too much? In many areas of science only ten or twenty really important papers appear in one year out of several hundred "relevant" ones published. The research worker in the field may have to choose whether to identify and assimilate them and then apply them to profitable research, or to spend his time studying a mass of literature which is of dubious or negligible value to what he wishes to do, even though formally speaking, it lies within his field. In this sense, the over-informed chemist, like the overloaded time-sharing computer system, may find himself so completely occupied with this bookkeeping activities that he has no time left for useful work. It seems entirely probable, since chemical information experts have now developed good systems for current awareness, that they will turn their attention to developing much more selective methods for the dissemination of information than merely sending the recipient everything that matches his interest profile. At the present writing, the principal price a scientist pays for being over-informed is an unpleasant feeling of being buried alive and having his time wasted through the evaluation of useless output. In the future, if the price of royalties for alerts is increased rapidly, the cash charge for unwanted or unused information may become one of the highest prices that has to be paid for these new services.

The efforts to design new chemical information systems did not start without a clear mandate from the chemists, who complained for years about their inability to cope with their expanding primary and abstract literature, as well as the increasing unwieldiness of their traditional indexes.

While there is no defense for the pseudo-information scientist who spends his time promoting jerry-rigged systems, there is correspondingly a clear rationale for the support of competent information scientists who demand the opportunity to use the scientific method in their explorations.

The chemist is annoyed when the layman does not see the relevance of the study of atomic structure to organic synthesis; so is the information scientist angered when the chemist does not see the relevance of his basic research in linguistics, semantics, logic, and behavior to the design of information systems.

So perhaps the chemist must retreat from his position of wanting to deal only with the black box, and acquaint himself to some extent with phenomena that influence information transfer. Otherwise, he would either have to submit passively to whatever is provided to him, or turn his back to developments that should be influenced by his insights.

As the information scientist matures in his understanding of chemists, and as chemists understand better the dilemmas faced by the information scientist, it becomes clear that in many cases the information scientist erred as much in exaggerating the value of the newly-developed alerting systems as the chemist did in exaggerating the efficiency of the invisible college for keeping him informed.

VI. EDUCATIONAL IMPLICATIONS

The role that computer-based chemical information services should play in the university education of chemists is a moot point. In view of the cool reception such services have received so far in the United States and the United Kingdom, it is even more questionable what role they will play in the near future. The universities and colleges should not only be one of the largest markets for computer-based chemical information because of their many and varied research programs, but because they are also the training grounds for the scientists of the future.

It is often argued among information specialists that the universities should be taking a leading role in the instruction of the coming generation of chemists in the use of computer-based services. However, as suggested above, the academic community, both in the United States and the United Kingdom, has not accepted computer-based services offered to them with the hoped-for enthusiasm, even when they have been provided free of charge. The phase-over to paying for those services at Pittsburgh, Georgia, Illinois Institute of Technology Research Institute, and Nottingham has been far short of the expectations one might have anticipated from those recipients who said they were "hooked" on free services. We have already considered to some extent the initial failure of the academic market. Let us now consider in perspective the value of introducing instruction in computer-based information services into the chemistry curriculum, either formally or otherwise.

It is probably true that the present generation of graduate students has received less formal instruction in the use of chemical literature or library services than any previous one since the nineteenth century. This fact is not lost on scientific librarians or chemical literature experts, who hasten to point out that the erosion of instruction in their area has occurred most unhappily at the very time when the chemical literature was going through its greatest era of expansion. Put together, this suggests an alarming picture of the development of an enormous literature in chemistry which no one will be able to use.

In our opinion, the growth of the scientific literature and the lack of instruction in its use, are directly coupled to each other and are not, for the most part, a serious misfortune. As chemistry has expanded, the scientific curricula have been strained to the point where many important areas cannot be touched at the undergraduate or even the graduate levels. Since World War II the pressure and pace in most university chemistry departments of any note have reached a level where the competition for attention and people's time would be almost unbelievable to anyone who had not been exposed recently to a first-class university chemistry graduate school. The fact is that many of the chemistry literature courses which were provided in earlier generations were taught by some of the least active practitioners on the staff. Although there were a few important exceptions, they often wasted the valuable time of students and the students knew it. Long hours were spent in learning to use the Beilstein system for locating compounds in the famous handbook of organic chemistry. A wise student quickly found that he was able to find them when he needed to, by means of the indexes, thereby saving himself many hours of studying an area which would bear professional fruit only if he became a chemical literature specialist.

The fact is that university chemistry professors and most of their graduate students learn rather quickly to use <u>Chemical Abstracts</u> and the major literature works of chemistry well enough for their own purposes. If they do not learn to do it perfectly, the amount of time which they lose through their lack of formal training is small compared to the amount of time that they save through avoiding lengthy courses on chemical literature and information.

The graduate student who is joining a large industrial concern probably knows in advance that the type of organization for which he will do research probably has excellent library facilities with trained personnel who know how to use information storage and retrieval systems and are capable of preparing suitable retrospective and current-awareness profiles for him. The chemist who looks forward to an academic career probably has confidence that when the chemical literature must be handled by special new techniques, he will either be able to learn them quickly or information specialists in his library will handle them for him. We are convinced that the most important ninety per cent of what a chemistry graduate student should know about the use of the library and the special resources that are available in it can be presented in three to six one-hour lectures. What place, then, is there for specialized training in chemical literature or computer-based services in modern colleges and universities?

There is room for a small number of departments which may choose
to specialize in the training of first-class candidates who will go with good
qualifications to some of the best chemistry libraries in the country.   Just
as Brooklyn Polytechnic Institute and one or two other universities have
chosen to provide special training in polymer chemistry, there is room for
several universities to provide special and thorough training in chemical
library and information science as part of their graduate curricula.   Other
departments might offer a small number of orientation lectures on the
chemical literature to all graduate students, which would be a useful part
of the first week or two of general orientation in any graduate school.

As computer-assisted instruction through interactive systems becomes
more common, we may expect to see a greater interest in the storage and
retrieval of chemical information in university computer centers.   In this
way computers will assist in providing rapid answers to questions that are
posed to them in the same way that automation in the laboratory has drast-
ically reduced the time required to obtain the answers to many experimental
questions.   This suggests that much more emphasis should be placed on
training students to ask questions, since so much less of their time as professionals will be required in developing strategies for obtaining the answers.   It may be a frightening thought to some, and a stimulating one to
others, to consider that we are working toward the day when the answer to
many of the scientific questions that previously would have required three
years of doctoral research may be obtainable in the laboratory within a
week, or obtainable within a day through the use of information stored in
the university computer center.   If this is to be the pattern for the future,
our educational practices in the present must be geared to it.

Chapter 2

THE PITTSBURGH CHEMICAL INFORMATION CENTER -
INTERNAL AND EXTERNAL INTERACTIONS

Edward McC. Arnett and Allen Kent

Department of Chemistry
University of Pittsburgh
Pittsburgh, Pennsylvania

and

Graduate School of Library and Information Sciences
Office of Communications Programs
University of Pittsburgh
Pittsburgh, Pennsylvania

I. BACKGROUND AND DEVELOPMENT OF THE PROJECT

From its inception the Pittsburgh Chemical Information Center has had two fundamental goals, which are reflected in its organization, its activities, and in the organization of this book. The first of these was to establish an Experiment Station for Chemical Information, where new services could be presented and tested with a wide variety of users. The second aim was the development of a viable regional information center, with a sufficient mass of facilities and capable personnel to disseminate those services which proved successful in the experiment station to a broad market of users on a cost recovery basis.

Although a number of challenging technical problems in systems development had to be faced in the course of the project (and are reported in later chapters), we believe that its most novel and important feature has been the behavioral study of the reactions of the user group. It was realized from the outset that we had a golden opportunity to conduct a well-planned evaluation of the reactions of an unusually large group of scientists to new information services. Properly conceived, this study need not be limited (as so many other studies have been) to small groups, who were studied by one or two behaviorial methods, after the fact, or at one point in time. A rigorous, broad-scale longitudinal study would be possible which would not only be of considerable academic interest but which might help to stimulate the orderly growth of a national and international system of enormous potential value. Clearly, the development of expensive services for a nonexistent or unreceptive market would be tragic, and we have believed from the start that the responsible accomplishment of our aims has depended on a behavioral program that was planned and executed carefully.

Since we have given top priority to the user and his reactions in the design of the project, we have also felt that this was the most interesting aspect of our study. Accordingly, we have emphasized the behavioral aspect of the project by locating the report of the Behavioral Group toward the beginning of the book (Chapter 3). Furthermore, the relationship of the other parts of the project to the behavioral effort is emphasized throughout the book as each of the associated task groups in the Computer Center, Library, and Programming review their experiences.

The Pittsburgh Chemical Information Center is unusual in being designed and operated primarily by chemists for chemists. The University of Pittsburgh Chemistry Department has played a key role in the development of the project. The Department, through its Library Committee, had taken a strong interest for some years in developing its Library to the most advanced stage possible. At the time the initial grant was received, the Department had already implemented Chemical Titles searches through a specially developed program accessible from a console in the Chemistry

Library to the University time-sharing computer system. This development and a number of related ones at that time were financed by the University's own funds. The Department, therefore, was greatly interested in plans to establish a Chemical Information Center, and it has given unstinting cooperation and enthusiastic support to the project. This has been evidenced in a number of ways, including, for example, the appointment of a full-time regular staff member in computer applications to chemistry and the establishment of an advanced graduate course in Chemical Information. However, its greatest service has been the day-by-day contributions of useful feedback information and advice by many people within the Department.

It was realized from the outset that the University of Pittsburgh and the tri-state area surrounding Pittsburgh provide a unique milieu for the development of a Chemical Information Center. The University has had, over the last decade, an increasing capability in knowledge availability systems, and its interest in this area was represented very clearly in the background of three of the coinvestigators. *

As the plans for the project developed and we became increasingly aware of the important role that behavioral science should play, we were pleased to discover sociologists on the campus whose interest and expertise in the computer manipulation and evaluation of behavioral records contributed to the development of a strong research program in this area.

Of equal importance to the existence of capabilities within the University structure was the realization that a large and varied group of scientists interested in chemistry existed within walking distance of the University of Pittsburgh Chemistry Department, in the Oakland area of Pittsburgh. A rough census indicated that nearly one thousand such scientists, covering a broad spectrum of interests, could be found in the University of Pittsburgh Chemistry Department and associated departments, at Mellon Institute, Carnegie-Mellon University, Duquesne University, and a variety of health-related institutions within a mile of the University of Pittsburgh.

As the project developed, users in these various institutions served as valuable control groups to check on possible biases due to idiosyncrasies which might exist in their populations as a result of field of interest or type of institution. Later, they were joined by other groups and individuals at carefully chosen university, industrial, and government laboratories outside the Pittsburgh area.

---

*Allen Kent, Director of the Knowledge Availability Systems Center; Orrin Taulbee, Director of the Computer Center; and C. Walter Stone, former Director of University Libraries.

In light of this historical development and a deliberate orientation toward taking advantage of it, the project has had certain corollary characteristics which bear comment.

Its user-orientation has caused us to put principal emphasis on those computer-based information services which are ready for presentation to the chemical user. It has therefore been a matter of policy to avoid doing research on systems development, computer programming, or behavioral analysis, unless they are related directly to the basic mission of the project.

The project has had a high degree of interdisciplinary complexity from the start. Since our goal was to study the methods of presentation and the acceptance of computer-based chemical information services, a broad range of facilities and highly-trained professional personnel were required in the design and evaluation of the experiments (Fig. 1). Magnetic tapes were received from the suppliers, and programs were implemented in the Computer Center by the programming groups. Searches for the chemical users were submitted to the Chemistry Library in the context of an experimental design developed in the Knowledge Availability Systems Center and the Behavioral Research Group, who also were responsible for evaluation and data reduction of feedback information.

Although a number of participants in the various task groups have had training or experience in chemistry or with chemical files, a major problem in starting the project was developing the internal communications

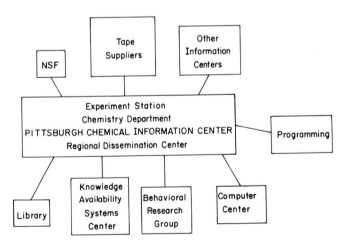

FIG. 1. Organization chart of Pittsburgh Chemical Information Center.

necessary for providing a meaningful organization in the initial phases and the type of mutual understanding required for good coordination as work proceeded.

From the vantage point of the principal investigator, these problems of organization and communication presented the most serious and demanding difficulties in getting the project underway. They were also the source of considerable satisfaction as they were surmounted.

A. Administration of the Project

When an information retrieval experiment is conducted in vivo, much less control of variables is possible than in an artificial situation. The experiment conducted with human subjects becomes very complex when many interactions are involved, including not only subjects, but also the other people who make up the environment.

Responsibility for development and operation of the Chemical Information Center was vested in five units of the University of Pittsburgh, shown in Fig. 1 and represented in various ways in the chapters which follow. The five units involved were the Chemistry Department, the Chemistry Library, the Computer Center, the Knowledge Availability Systems Center, and the Behaviorial Research Group. It was believed that these groups would constitute an effective nucleus of optimum size to execute the program.

In addition to the principal investigator's responsibility for coordinating the project, he was closely involved as a research chemist in the actual exploitation of the system at the operating level—indeed, this was his chief motivation for initiating the project in the first place.

Coordination was necessary among the following participants within Oakland and outside Pittsburgh:

1. Within the Project—The principal and coinvestigators kept in close touch through frequent meetings at least once a month. Coinvestigators were responsible for the contributions of their task groups. Generally, a project director was expected to handle the operational details of his task group assignment.

2. Within the test area—As various user groups were to be identified and as various libraries outside the University Chemistry Library were to gain access to the Chemical Information Center, it was necessary to develop communications with them. This was done by the principal investigator and the director of the Chemistry Library Task Group.

3. With Chemical Abstracts Service (CAS) and National Science Foundation (NSF)—Contact was maintained with NSF and CAS through meetings of the appropriate participants.

4. <u>With outside suppliers of chemical information</u>—Access to and col-
laboration with other documentation and data centers were expected to develop and did. The group looked forward to integrating with regional, national, and international networks. The principal investigator and appropriate
coinvestigators were to coordinate the development of these relationships.

In addition, the Chemistry Department committed itself to securing the
appointment of a staff position in computer applications to chemistry which
could lead to tenured rank.

B. Tasks Proposed, Completed, and Dropped

This organization was assembled in order to meet the goals of the project. We will now consider in greater detail how these aims were to be
approached through specific tasks assigned to the different task groups.
In order that these efforts be coordinated and so that cost estimates could
be made for completing them, a task schedule was developed. It will be
presented here, since it can serve as orientation for the development of
the projects described in the chapters by the separate task groups. It may
also be of value to readers who are interested in organizing similar efforts.
Most of the remainder of the book will be devoted to reporting on the completion of a number of the tasks originally proposed. It is important to
recognize that for a variety of reasons (economic, technical, and judgmen-
tal), some of the tasks which seemed important in the original planning
were dropped as the project actually developed. We now turn to the original proposal (March 1968) in order to describe the tasks as they were projected at that time.

C. General Description of Tasks

The aim of the Chemical Information Center is to provide chemical
information services in a controlled and systematic way to a typical group
of users and thereby provide <u>information</u> about the <u>use</u> of chemical information. This implies two types of tasks from the very start of the project.
1. The development of technical facilities and personnel to develop
the physical installation and accessing procedures for presenting new services to the users.
2. The development of background information and operating statistics
about the users and their activity in the system so that there will be a
corpus of reliable facts amenable to objective analysis at various future
points.

Although the complete development of an experimental station might well encompass a period of ten years, intermediate goals must be set to assure rapid progress during the first three. The first phase of the program would involve the following activites:

1. Acquisition of Materials. This would require the physical acquisition of computer-based files or development of direct access to some held in other locations. The Chemical Abstracts materials would, of course, represent the majority of such sources. However, access to services offered by other organizations would be required in order to permit comparison. Available programs would be checked and new ones written as required for the direct use of files or by remote access. Arrangements for supplying hard copy of source materials on demand would be made.

2. Development of Data Collection Mechanisms. Information on usership would be collected by computer, by questionnaires, and by interviews. Obviously, an appropriate balance must be struck between harassment of users with questionnaires which might inhibit their use of the center, and a complete failure to solicit feedback information from them by established scientific procedures. Accordingly, we would make every effort to see that the computer itself keeps track of the usership as far as is practical. We would also use a Behavioral Research Group to develop a research design for obtaining information from users and to devise strategies in the use of questionnaires and direct personal interviews for supplementing the data which are accumulated in the computer itself. Although chemistry will be the only testing field of this project, long range implications of the study for other disciplines will be kept in mind from the start.

3. The User Group. The thousand-odd local potential users of chemical information shown in Table 1 would have most direct access to the Chemical Information Center and would most naturally be its heaviest users. However, other chemists in the greater Pittsburgh and tri-state area would soon wish to take advantage of the Center.

An initial questionnnaire now being prepared by the Behavioral Research Group will soon be distributed to chemists in the area so that background on the unperturbed user group will be available for comparison in later follow-up surveys. Users of the Chemical Information sources will have to provide some vital statistics before obtaining an identification number for computer use. This will accompany every use they make of the system.

We will therefore develop rather quickly an extensive file of statistics, profiles, questions, and user responses from which pilot groups of especially enthusiastic users might be drawn for testing specific new data bases.

TABLE 1

Potential Users of Chemical Information in Oakland Area of Pittsburgh

	Total	Administrative	Analytic	Biochemistry	Chem. Eng.	General	Inorganic	Metallurgy	Nuclear	Organic	Physical	Crystallography
Bureau of Mines	87					87						
Carnegie-Mellon												
Chem. - Staff	18	1					3		3	4	6	
Students	83			1			14		17	30	22	
Chem. Eng. Staff	10				10							
Grad.	65				65							
Undergrad.	103				103							
Mellon Institute												
Staff	10	1						2	1		6	
Senior Fellow	48		6	11	1		2	1		12	15	
Fellows	86		15	5	1	3	2	5	7	24	24	
U. of Pittsburgh												
Biology-Staff	5			5								
Chem. - Staff	29	1	5				3		1	9	10	
Post-doc.	22		1				4			6	11	
Grad.	129		9	2		43	14		5	31	25	
Undergrad.	127					127						
Engineering												
Staff	22				11			11				

Category		No.											Total
Students													
Undergrad.					164		113			50				227
Grad.					65		54							119
GSPH – Staff		12								5				17
Students		41								11				52
Medicine – Staff		9												9
Students		27												27
Pharmacology														
Staff		4												4
Students		4												4
Crystallogrphy														
Staff	14													14
Students	13													13
Total	27	121	36	3	420	260	186	42	116	50	119			1,380
Westinghouse Res. Lab.														139
Duquesne U. Chem. Dept.														
Staff														8
Students														30
Carnegie Public Library – No figures available														
Local hospitals – 20–50 active chemists														

November, 1967

It is part of our understanding with Chemical Abstracts that there will be no legal obstacle to the free use (without charge) of any service in the Center as long as it is in an experimental mode. This will normally be twelve months from the time it has been made available to the usership on a regular working basis in the searching mode under study.

4. Development of Orientation and Training Procedures. In order to assure that these will be administered under controlled conditions, formal orientation and training mechanisms will be developed involving both computer-assisted tutorials and videotaped orientation sessions.

5. Specification and Ordering of Equipment. Although the basic equipment and facilities for developing the experimental station are available at the University, some additional equipment will be required, such as new terminals to the shared-time computer, and a dedicated mass store. Specifications for such equipment would be developed and orders placed.

6. Development of Accessing Procedures. At least two accessing procedures to files may be foreseen: 1) direct access by users to local and remote files; and 2) access via an intermediary, either face-to-face or by telephone. The latter is needed for infrequent or casual users who do not have nearby terminal access. In the case of direct access, the training program (section 4 above) would be used.

7. Development of Instruments for Collection of Data About Usership. Based on the mechanisms developed in section 2 above, information about the usership would be collected. The depth to which information on the response of users to the system could penetrate would depend largely on the level of funding. The various levels of user analysis may be separated crudely into the following categories and will be described in some detail under the task of the Behavioral Research Group:

(a) Bookkeeping by the computer of the user's activity with the computer-based services. The value of getting these data without preliminary knowledge of the user's normal use of the chemical literature or any feedback of his reactions to the new services would be small.

(b) Uncritical evaluation of user's behavior and reactions by computer and clerical personnel.

(c) Analysis in depth of how and why different types of chemists use different chemical information services and the effect that it has on their literature searching habits and professional effectiveness.

8. Development of Mechanisms for Continuing Communication(s) with the User Group. A newsletter or other communication medium would be developed to alert users quickly regarding the introduction of new pilot services available to them and to inform them on the progress of experimental programs.

9. Data Reduction and Reporting. The design of statistical analysis procedures and reporting mechanisms would be developed in order to assure rapid feedback of experimental results both to NSF and to CAS.

D. Succeeding Phases

During the first phase of the program a number of simplifying assumptions would have been made in order to develop the experimental program as rapidly as possible, consistent with sound design. However, during the succeeding phases additional problems could be explored, as follows:

1. Conversion from Batch to Real-Time Searching

A limiting factor on the first phase will be the necessity of handling a number of the operations by batch processing until a dedicated disk storage and/or data cell capable of holding a large quantity of information for searching via terminals to the shared-time computer becomes available. We believe that an important obstacle to the use of computer-based materials by research scientists is the delay between question and response in batch searches. Accordingly, procedures and programs would be developed to permit direct access to files in real time.

2. Augmentation of Available Data Banks

During the first phase, the personnel of the center will be learning the techniques of introducing users to the different services; they will be learning to use new experimental materials themselves and will be discovering which types of materials that are held in centers outside of the Pittsburgh area are in greatest demand by the user group.

E. General Task Schedule

This has been developed on the basis of six-month increments, since this seems to combine a series of fairly natural stages with the budgetary cycle.

We will expect to file quarterly as well as annual reports.

The first six months (January 1—June 30, 1968) would be spent in acquiring materials, deploying operating personnel, programming new services for use in our computer, identifying members of the general user group on the basis of the first questionnaires and interviews, the development of improved questionnaires and distribution to a larger sample of potential users, and working out the details of our relationships with Chemical Abstracts and the National Science Foundation as well as the purveyors of any other materials which we would wish to use. During the period, we will be "setting up our shop" and developing the administrative and evaluative procedures. We will not be hitting our stride until the end of this period; and, therefore, the budgetary demand will be smaller than when we are completely staffed and handling materials on a regular experimental basis for a large group of users.

From July 1 to December 31, 1968, we will be following through on the first phase of the project, during which we are operating primarily with batch processing. The services and procedures which have been started during the preliminary six month period will now be presented to a larger group of users. At the same time, we will be making plans and perhaps doing some reprogramming for the time in early 1969 when our dedicated disc arrives and we can make our first attempts at phasing some large elements of the project over onto real-time interactive mode. Some experience will be gained by interactive programs during the first six months with small files presently available in the Chemistry Library. Tapes will be used and also rented space on an eight-disk pack scheduled for delivery in November 1967.

Some time between January 1 and June 30, 1969, the dedicated disc and associated equipment for the Chemical Information Center should arrive. We should expect that period to be spent in continuing the normal operation of the first phase of the project, while at the same time the equipment and other changes that will be involved in going to real time are being instituted. New questionnaires will be developed during this period and new programs for keeping track of users in real time operation will have to be prepared.

By July 1, 1969, we should be operating in real time on a regular basis and should be attempting to phase several of the experimental services into it. Very likely, a number of materials which were tested first on a batch mode will be examined again experimentally for retrospective searching when, and if, they become available in an interactive mode through the console. Other services will more properly be treated batchwise throughout. By this time, a number of our materials which were previously available on an experimental basis will have been completely phased over to operational mode. During this period we expect that some of the most intensive work on examining the registry system in an interactive mode will begin and the first experiments on interactive retrospective searching will be taking place.

During 1970, the project should have entered the "steady-state phase," where new experiments are being introduced, some in batch time, some in in real time, and old materials are being phased over to become operational services available through some fee mechanism to anyone in the region who wishes them. During this period, the entire project will be evaluated and plans made for the next stage in the development of the chemical information center.

The projected schedule for implementing various services in different modes of operating is shown in Fig. 2—in which the various services are represented as follows:

Condensates - CA Condensates, the service which eventually became the principal subject of study at the Center.

ASCA - See section on ISI.

IR - NMR - Infrared and nuclear magnetic data files.

Crystallography Files - A collection of crystallography data accumulated in the Crystallography Laboratory and considered for development into an interactive data search system.

CBAC - Chemical-Biological Activities.

MEDLARS - Medical Literature Analysis and Retrieval System.

CT - Chemical Titles - The tape service from Chemical Abstracts Services. Originally presented in interactive mode and later studied intensively as a batch mode as a current-awareness service.

POST - Polymer Science and Technology.

BJA - Basic Journal Abstracts.

SCI - Science Citation Index, a hard copy search tool from ISI.

DATS - A hard copy Desk-Top Analytical Tool for searching part of the common data base produced by CAS.

Structure Searching - Experimentation with the CAS Structure Registry system was considered to be an important aim. In preparation for this, three members of the project, two chemists and the programming manager, participated in a workshop at CAS during late May and early June of 1969, and the CAS structure search system was implemented in the computer center. Prior to this, preliminary exposure of a group of graduate students to structure searching was made through the Desk-Top Analytical Tool. Custom searches were prepared at Dr. Lefkovitz's group in Philadelphia on compounds from the common data base and a comparable group of compounds were searched at ISI using the ICRS system.

The accomplishments of the above proposed task schedule were dealt a severe blow in the fall of 1968, when the first shock of the funding crisis on NSF grants struck home. The project emerged from this experience with sharply curtailed budget and plans. A year after start-up a number of the originally proposed plans were clearly behind schedule or had been abandoned and we could make the following statements in the first annual report.

"1. Our plan for an extensive behavioral study using repeated depth interviews of many users has been a victim of the budget crisis. We have decided to settle for descriptive behavioral data rather than that which is amenable to fundamental interpretations.

2. We have made no attempt so far to introduce BJA, POST, or CBAC, both because of a lack of user demand and in order to prevent overtaxing our resources. * The search system for the Preston NMR file has also been dropped through lack of general supporting interest.

*Subsequent experience at Nottingham vide infra indicates that this was a good decision.

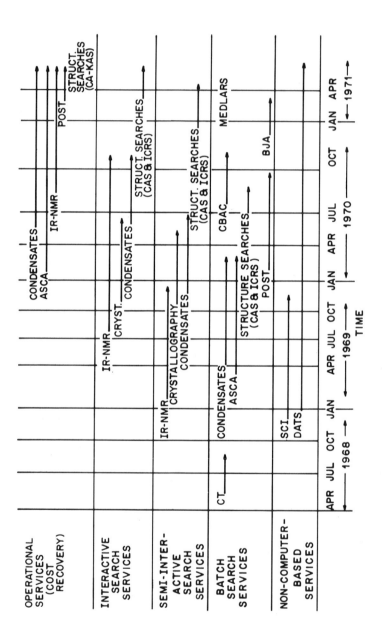

FIG. 2. Projected schedule for implementation of services.

3. We have made very little effort so far to develop manuals and teaching aids for profiling. This is chiefly because those which have been developed at CAS and Nottingham satisfy our needs for the workshop presentation which we have used so far.

4. Our experience with running Chemical Titles in the Pitt Time Sharing System has dampened our enthusiasm for rapid development of large experiments in the semi-interactive or interactive mode. With limited resources we are instead trying to get our batch processing costs as low as possible."

Shortly thereafter it was decided to drop our plans to implement the CAS Registry and structure search system. This decision was prompted by several considerations—the chief ones being the following:

1. The infeasibility of doing intensive substructure searching on most of the million— odd compounds then in the system, since there was little likelihood that they would be suitably fragmented for such searches in the foreseeable future.

2. The enormous commitment of facilities and personnel that would be needed to use the system.

3. A general lack of interest within the Chemistry Department in using the file in view of number 1 above. This was true despite several seminars in which the structure-registry system was described in detail by the group that attended the seminar—workshop at CAS. We were generally unable to answer in credible terms the question posed by our users, "For what purpose would we use the present structure-registry system?"

4. The University of Georgia had already taken considerable initiative in implementing the system and was experimenting with it.

Accordingly, we felt that as a user-oriented center we should throw our chief effort into experimentation with current-awareness and retrospective searching (in which our users expressed major interest) and purchase structure searches from Georgia if any of our users wanted them.

With the exception of some experimentation with special small files and interactive systems, the project after the first year became dedicated to doing the best possible job of implementing CA Condensates searches, in both retrospective and current-awareness modes, for a large and representative group of users.

F. Interactions During the Conduct of the Project

Once the project was funded by the National Science Foundation, it was, of course, necessary to bring to bear on the tasks at hand the staff and facilities identified at the start.

First was the acquisition of the data bases, the most important one being <u>Chemical Abstracts Condensates</u>. This came later than expected, but not exasperatingly so. It was expected that the program package provided by the vendor would have to be redone if it were to be processed in a cost-effective manner. An ad hoc working group was established to develop specifications for a new computer program. This group consisted of members of several units of the University.

The work of this group was interrupted by the news that IBM had developed the TEXT-PAC program, and, based on a preliminary evaluation, it was decided to devote considerable effort to the implementation of this package rather than to undertake a massive program-writing job of our own.

As discussed in Chapter 4, this decision was made because 1) TEXT-PAC had purported capabilities for efficiency and effectiveness that surpassed our minimum specifications, permitting us to consider cost-effective retrospective searches; and 2) the program was claimed to be in working order.

Another factor that influenced our decision was the underlying thought that if the package were as good as it seemed, it might be adopted by other tape suppliers and possibly lead to the development of standards.

The next major problem was the development of feedback routines, which would assure the receipt of relevance judgments from users in a reliable and consistent fashion. Those that were finally accepted are described in Chapter 3.

Following this came the tribulations of getting current-awareness services established on a regular schedule, and, when this had been resolved, of developing programs to permit output to be obtained in unit record format.

Many conversations were necessary with <u>Chemical Abstracts</u> personnel, since they, too, were still struggling to provide error-free tapes on a regular basis.

Discussions with many other groups were required as the user community was expanded to include other departments in the University, other academic institutions, and chemists from industry and governmental organizations who were willing to participate in the experimental program. The workshop was the mechanism chosen for this purpose.

G. Contacts with Other American Information Centers

It was recognized soon after the start of the project that the development of the Pittsburgh Chemical Information Center was closely related to the growth of a number of other emerging information centers, some of which in due time might bécome part of a national or international network. It was expected that they would face many common problems and that they would soon need a forum to discuss them. Furthermore, since most of the data bases used by the information centers would be prepared by outside tape suppliers, the information centers would need an organization to provide a common voice for expressing their needs to the tape suppliers. Accordingly, during the winter of 1968-69, a small group of representatives of the emerging scientific information centers met to discuss problems of tape format, user reactions, marketing, and a variety of related questions. The need for organization was so strongly felt that the group quickly gained momentum and was formalized into ASIDIC, Association of Scientific Information Dissemination Centers, which at the time of writing was comprised of 20 members and 42 associate members. We regard the participation of PCIC in the founding and development of ASIDIC as an important and interesting activity of the Center.

H. Contacts with European Information Centers

About a year before the initiation of PCIC, a Chemical Information Center was established at the University of Nottingham, England, under the direction of Professor A. K. Kent, a biology professor with strong interests in computer systems. Professor Kent's contacts with Chemical Abstract Services and the American chemical information centers in Pittsburgh, the University of Georgia, and IITRI have initiated a valuable international exchange of information about problems in the use of computer-based chemical information. Meanwhile, other European information centers had begun programs for the dissemination of a variety of services. They, like their American counterparts, banded together to form a user organization EUSIDIC (European Association of Scientific Information Dissemination Centers) now comprised of the following organizations: United Kingdom Chemical Information Service (UKCIS), England; Biomedical Documentation Centre (BMCD), Sweden; Royal Institute of Technology Library (KTHB), Sweden; Danish Technical Library (DTB), Denmark; Institut Français de Petrôle (IFP), France; Netherlands Organization for Chemical Information (NOCI), the Netherlands; Shell, England/the Netherlands; and Unilever (Research Division), England/the Netherlands.

During the summer of 1970, the principal investigator served as Visiting Professor of Chemistry at the University of Kent in Canterbury. This visit provided a number of occasions to compare the British chemical information program with our own. In addition to a number of discussions with members of the faculty and student body at Canterbury, there were many chances to talk with a number of British academicians during an unusually large conference on reaction mechanisms, which was held at the end of July. This covered the areas of organic, physical, inorganic, and biological chemistry and was attended by more than 400 chemists, 60% of whom were British. Conversations were also held with scientists of Unilever Corporation, one of the largest British chemical firms, and with Dr. A. K. Kent and his assistant, Dr. Kabi, who were principally responsible for the United Kingdom Chemical Information Service (UKCIS) at Nottingham.

The Nottingham Center managed a large and very well-organized experiment in the United Kingdom, in which free service based on Chemical Titles (and in some instances CBAC and POST) was offered to almost any faculty member, postdoctoral, or graduate student who wanted it. Nearly a thousand students and staff participated and were carefully instructed in the use of the services through liaison officers who visited their departments. A large number of industrial organizations were also instructed in the use of the services and were eventually required to pay for them.

The academic groups were warned for some time in advance that free service would come to an end. However, quite a furor was raised when it was finally announced that as of July 1, 1970, academic users would be expected to pay for their profiles. In view of the rather feeble reception of computer-based chemical information on a pay-as-you-go basis in the United States, it is interesting to compare reactions in the United Kingdom, where a number of the variables in the situation are quite different. For example, English universities had been much better prepared for computer-based chemical information than American ones and were not in the midst of the financial emergency which had reduced many American departments and scientists to the point of panic about sources of funds. Against this background, it was interesting to find that, in general, the observations which we had tentatively made about the poor acceptance of chemical information in American universities as compared to American industry appeared to carry over into the British scene at this time.

British industrial chemists from Unilever were very enthusiastic about chemical information services and had learned to take them almost for granted. On the other hand, many conversations with English academic chemists made it clear that they (like their American counterparts) were often enthusiastic in principle about chemical information services but did not think that they were valuable enough to spend the $150 to $200 from their research funds which a suitable profile would require. This difficulty was recognized by the people at the Nottingham Center, and they had started to do something about it both in terms of a marketing approach and also at the political level in dealing with British granting agencies. We will return to this matter later, but at this point it may be more suitable to document some of the attitudes expressed in conversations.

At Oxford, where there is considerable experimentation on the acceptance of computer-based chemical information in England, about 60% of the organic division had participated in the experiment, and about 50% of the group wished to continue receiving the service but were not prepared to pay for it from their own grants. The physical chemists at Oxford were quite divided about the value of the service and preferred to put their money elsewhere. At the time of this writing there appeared to be little or no interest in the system at Cambridge. At Sheffield, where a major study on structure-searching was being carried out by a staff member of the library, there was little appreciation of his effort or interest in computer-based services within the Chemistry Department.

In general, English academic chemists did not know about CASCON, their experience being limited entirely to Chemical Titles, the shortcomings of which they were quick to point out.

The feeble reception of the UKCIS products by academia was well known to the Nottingham group. Their marketing approach in dealing with the situation was to offer profiles, for the coming year only, to academic users at half price, that is to say, about £50 per profile for the year. This was possible because computer time was contributed at a very low cost by the Oxford Computer Center. After that, money would have to be found from other sources to pay full price for the service. By the summer of 1970, a half-price offer had brought favorable responses from about 120 of 600 academic users who were enthusiastic about the service they had previously obtained for nothing. The Nottingham group also offered a low-cost introductory service at £25 to other users. It was expected that these users would be almost sure to continue with it at the regular price once they learned how to use it. This suggested that if a self-selected group of academic chemists would purchase information services at a minimum introductory cost, they would probably continue with the service provided a means were found to get money to them without too much sacrifice of their research budgets. It has been assumed, on both sides of the Atlantic, that the academic community is potentially one of the biggest and most important markets for chemical information and that the basic problem is to find a mechanism for supplying grant funds to support it.

In the United Kingdom academic research is supported in two ways: one is through the University Grants Committee, an agency which supplies massive grants to the universities and these in turn are distributed within the university to libraries, departments, and so forth. Diversion of these funds to the support of chemical information services obviously involves head-on competition between the library and the chemistry department, or between those who want the services within the department and those who do not. Since the University Grants Committee does not instruct the universities on the allocation of their grant funds, there is no reason to suppose—If these grants were increased—that any of the extra money would flow into information services.

In the first instance, British libraries (like many American ones) consider that these are new services which they have not supplied before; and, since their budgets are already inadequate, they hesitate to cut into them for the support of this type of new and potentially expensive service. The struggle for funds, therefore, takes place at the departmental level where money for profiles must compete against money for instruments or for supplies. Instruments and chemicals are generally regarded as of primary importance to research, and information services are considered (at best) secondary. Thus, the proponents of chemical information (who rarely make up more than a third of the department) are outvoted, and the issue goes by the board on an all-or-none basis.

Scientific research in the United Kingdom is also supported through individual research grants, which are similar in most ways to those provided in the United States. Here an effort is being made to see that grant applicants will first be permitted and then encouraged to apply for funds for chemical information as a line item of their budget similar to that requested for equipment or supplies. This raises an interesting question, since if the user feels he is operating within a tight budget, he may choose to leave out a request for chemical information services in order to save money. On the other hand, if he is told that he does not need to worry about requesting such funds, he is being told, in essence, that they will be continued on a free basis—in which case he is really not making any proper value judgment on such services.

The Office of Scientific and Technical Information (OSTI), which funded the Nottingham effort as an experiment, considered (much as the NSF would in the United States) that they could no longer support the project once it was operational. It seems only reasonable that political means will be used to find a mechanism for supporting the use of chemical information in the academic community. A large sum of money has been spent on experiments to introduce these services. It would seem foolish not to establish a way to supply the services to those scientists who find them valuable once their usefulness is demonstrated.

Although there are parallels between the situation in the United Kingdom and that in the United States, there are also some very important differences. In the United Kingdom, the experiment on chemical information in the academic community has not only nearly a year's longer run, but it has had the advantage of highly centralized and well coordinated planning and backing from the government, the Chemical Society, and the scientific establishment in general. Publicity in English journals has been excellent and a large number of younger English chemists have had some experience in using the services in the course of their university training. Nothing comparable to this has yet occurred in the United States.

As of August 1970, the Nottingham group was processing four data bases, and of these CASCON was the only one with a growing clientele. Chemical Titles was static. CBAC and POST were dwindling, with only thirteen paid profiles for POST. This has since been replaced by a special profile on the macromolecular section of CASCON. In total about 350 profiles were being handled, with about eight new paid ones added per month. This is very close to what was considered the economic break-even point for the operation. If one discounts the cost of research and development, which of course could be underwritten by appropriate research grants from OSTI, it is expected that all costs would be recovered and that the principal problems would arise from the growth of the operation being greater than planned. Accordingly, the procurement and training of personnel was becoming a serious problem. Nottingham was working closely with the Oxford Experimental Group, which was running comparisons on different services. They have since completed a comparison on the acceptance of Chemical Titles, and ASCA with about 300 users.

A major reprogramming effort was underway at Nottingham, including an interesting system for automatic profile development using the computer to study the terms which appear with greatest frequency in the relevant alerts returned by the user to the center. The computer then suggests a series of single terms which should be part of the profile, and these are examined by the user for rejection or inclusion in a modified profile. After three iterations the average profile could be brought up to about 90% recall relative to what the group estimates an ideal profile would bring in, and this is accomplished entirely automatically through use of the recycling system. About 200 users were in this experiment, and relevancy was determined on the basis of abstracts sent from Chemical Abstracts on Microfilm to accompany the titles. Naturally, the relevancy goes down as the recall goes up. A program was also developed which prints out a table for the user (on the basis of his profile and its previous performance) which tabulates the percent recall, percent relevance, and costs per year that relate to these performance levels. The user can then choose for himself what level of recall and relevance he is ready to pay for. This should aid considerably in marketing the services if the cost of royalties to the tape suppliers go up, since a person can see clearly what he may expect to pay for different levels of performance. Dr. Kent considered this approach a potentially valuable one in the development of bibliographies through retrospective searching.

For retrospective searching, a major effort was being made in the development of a new file-inversion program. This encompasses all the features of their present types of logic, including left-hand truncation.

The file-inversion system involves indexing through breaking words into fragments, which are then filed by means of the Key Letter in Context (KLIC) system.

J. Interactions with Industry

It was understood at the start of the project that the grant from NSF was to cover only experimental and developmental activities, and that once operational, the program would continue without Foundation support. Part of the understanding was that services to industry would be offered on a cost recovery basis. First there was the problem of determining costs, and an algorithm was developed in which the most important variable was the hypothesized quantity of profiles that would be processed in a batch. It was assumed that the cost per search would be reduced as the number of searches processed at one time increased. In order to keep fees as low as possible, it was necessary to develop a marketing program that would involve as large an industrial clientele as possible.

The marketing group of the Knowledge Availability Systems Center assumed this task, and was still developing marketing tools and engaging in sales efforts at the time this book went to press. Although this is an ongoing activity, it can be concluded that the only successful approach to cost recovery is selling the ability to solve problems rather than selling computer searches from any one of several chemically-oriented files. This is the case because the solution to problems of industry are not limited to the boundaries of a given file of information, no matter how extensive. Industry, therefore, seems more willing to pay fees for searches over as many relevant files as necessary to solve a problem, rather than to pay for the search of a given file against a profile.

II. INTERACTIONS WITH THE INSTITUTE FOR SCIENTIFIC INFORMATION (ISI)

For several years before PCIC was planned, a few members of the Chemistry Department at the University of Pittsburgh had been using the Automatic Science Citation Alerts (ASCA) current-awareness system produced by ISI in Philadelphia. Although this data base deals with a wide variety of physical and social sciences, it covers six or seven hundred of the key journals in chemistry, about the same number as are regularly handled by Chemical Titles. It is a very rapid and reliable current-awareness service. In addition to using key words and authors as search terms (as with CA Condensates and Chemical Titles), ASCA has a special feature of alerting the reader to references in which he is particularly interested. This entree to the literature is valuable in cases where the title to the paper gives no clue that a certain important area will be referred to within the paper, thus providing unusual penetration which may be lacking in key words and titles. It was believed from the outset that a complete current-awareness service should include citation searching, and plans were made to acquire the ASCA data base on tape and provide this as part of our own current-awareness and possibly retrospective search system. Some thought was given to a possible merging of the output from the different search services, such as ASCA, Condensates, and Chemical Titles, in order to reduce redundancy, although it was recognized that such merging could produce many problems.

When a serious reduction in funding for the project became imminent, it was decided to provide ASCA service to several user groups through purchase of the regular service from ISI rather than through processing of the tapes in Pittsburgh. This was done, and a study of the citation service was carried out with a special group of users, some of whom received CA Condensates and Chemical Titles. This project was supported principally by funds through the Pennsylvania Science and Engineering Foundation and will not be discussed in detail here. A much more extensive comparative study of ASCA with other services has been initiated at Oxford University in England.

A. Interactions with the University Administration

The consequences of success are sometimes more difficult to contemplate than those of failure. To start with the negative point of view, it would have been quite simple to complete a project whose conclusions were that no useful services could be derived from computerized chemical information files, to write a final report, and then go on to other matters.

Since it had been assumed that the services would be useful, it was necessary to consider, from the first, how to cope with success. The major problem is to assure continuing support of the operational services from means other than external funding agencies, since they are willing to support only research and development. Would the services be institutionalized so that regular budgets would be provided to permit continuing operations?

Here again the need for interaction was critical, this time with the University librarian and the administration. The Director of University Libraries was in continuing contact with the project, as a coinvestigator, and little more had to be done than to advise him in time that the service aspects of the project should be taken into account in the regular budgeting cycle.

The administration was kept informed periodically, so that they would be sensitive to the progress and therefore in a position to evaluate the budget requests made in relation to maintenance of operational services. This task was made easier by the development of a plan to provide similar services in other disciplinary areas on the campus-based information system.

B. The Campus-Based Information System

The reward for keeping the administration informed of progress and problems was an increasing interest on their part in expanding the data bases and services to users in other disciplines in the University and in the region. This interest has coalesced into a development program for a campus-based information system.* It also led to an administrative reorganization which would facilitate the coordination of this and related programs. An Office of Communications Programs was established, assuming responsibilities for the Computer Center, the University Libraries, and the Knowledge Availability Systems Center.

---

*This program has since been implemented under a grant from the National Science Foundation (G-27537).

Chapter 3

THE USER'S INFORMATION SYSTEM:
AN EVALUATIVE RESEARCH APPROACH

Daniel James Amick*

Pittsburgh Chemical Information Center
University of Pittsburgh
Pittsburgh, Pennsylvania

---

*Present address: Department of Sociology, University of Illinois at
Chicago Circle, Chicago, Illinois.

I. INTRODUCTION

This chapter is written primarily for behavioral and information scientists, because in recent years the work of the two has become intimately related. Behavioral scientists have become increasingly aware of the utility of information theoretical approaches to the study of behavioral phenomena and information scientists have exhibited a keen interest in the behavioral implications of information processing activities by humans. As a result, a creative symbiosis has developed. The behavioral scientist has discovered a new arena of human acitivity amenable for study and analysis, while the information scientist has found that he needs behavioral science methodology and theories in order to develop the necessary body of knowledge for the foundation of his discipline.

As behavioral scientists working on a project to establish an information center for chemists, we found ourselves performing dual roles. The first was to conduct evaluative research to determine user reactions to the varied information services made available to them by the project. The second was that of the behavioral scientist interested in the information processing activities of human beings. The first role is more concerned with methodological technique and straightforward reporting of the findings. The second role tries to determine the implications of the findings for behavioral science, information science, and information system design, and is therefore more discussion-oriented and more interpretative. By expressing both roles, we hope that the information scientist will see how such a study was performed, in terms of technique and design, and will be able to determine the implications of our findings for the design of his own information system.

The results of our work will hopefully have significance for scholars in several areas. Therefore, the general findings will be presented with enough detail to substantiate the main points. A detailed report of a finding might well interest a particular subaudience of readers but turn off many others. The details of particular studies have been and will continue to be reported in the periodic behavioral and information science literature.

To perform evaluative research on information systems, the investigation should have at least three primary objectives. First, it should describe the user population according to a classification scheme which best reflects characteristics which are relevant to a user's information needs and uses. This is a major step in giving system designers a "feel" for their clientele. Second, it should establish standardized mechanisms for systematically ascertaining feedback from users concerning their expectations of, and actual use of, the various information services. For example, in this project these feedback mechanisms ranged from general purpose depth interviews to special purpose feedback cards which were attached to the user's alerts. The third objective is to make certain that proper techniques of data gathering and analysis are employed so that the entire evaluative effort consistently exhibits reliability and validity. This requirement is paramount because an inappropriate methodology can cause more problems than it will solve. Therefore, we will emphasize technique. One important note should be added here. It has been our experience that once these techniques and methods have been devised, modified, and initially implemented, their administration can be turned over to other people with no training in behavorial investigation, i. e., librarians, information analysts, etc. However, the stages of conceptualization and design are clearly problems for the behavioral researcher.

II. DESCRIPTION OF THE USER POPULATION

In this section, we will describe some relevant characteristics of the user population: demographic, professional, and information-use characteristics of the group of experimental users of PCIC. We wanted our experimental user population to be representative of the various sectors of the Greater Pittsburgh community of chemists. These were: 1) large academic institutions, 2) small academic institutions, 3) large industrial organizations, 4) small industrial organizations, 5) a non-profit research institute, and 6) a governmental research agency. The distribution of the original user population is shown in Table 1.

The users from the University of Pittsburgh were both faculty (approximately 60%) and graduate students (approximately 40%). All of the Duquesne University users were faculty members and of the 25 users from Carnegie-Mellon, eight were graduate students. In the large industrial

TABLE 1

Organizational Affiliations of Users

Category and type	Number	Percent (approx.)
1. University of Pittsburgh		
a. Chemistry Dept.	72	31
b. Nonchemistry Dept.	33	14
2. Duquesne University	11	5
Carnegie-Mellon Universtiy	25[a]	11
3. Koppers Research Laboratory	11	5
Pittsburgh Plate Glass Res. Lab.	7	3
E.I. Dupont Res. Laboratory	7[b]	3
United States Steel Res. Lab.	10	4
Gulf Oil Corp. Res. Lab.	8	3
Westinghouse Research Lab.	3	1
4. Sinclair-Koppers	1	4
Mobil Chemical Co.	1	4
Blaw-Knox Chemical Inc.	2	4
Mobay Chemical Co.	1	4
Dravo Corp.	1	4
Matthey Bishop Inc.	1	4
Pressure Chemical Co.	1	4
Neville Chemical Co.	1	4
Benfield Corp.	1	4
5. Mellon Institute	21[a]	9
6. U. S. Bureau of Mines	11	5
National Science Foundation	5[b]	2
TOTAL	234	100

[a] Many of these users had joint appointments between Carnegie-Mellon University and Mellon Institute. They were classified according to the first-mentioned appointment.

[b] Though not located in the Pittsburgh area, these users were included in the original experimental group for special interest reasons.

group, the users came from diverse professional ranks: company research directors, section supervisors, research group managers, senior scientists, and junior scientists, many of whom were active bench chemists. The users from small industrial concerns had titles which, for any practical purpose, defy classification. All users from Mellon Institute had ranks of either Fellow or Senior Fellow. The users at the National Science Foundation were all program directors while the users at the U.S. Bureau of Mines were either research chemists or research engineers. In summary, the experimental user population was fairly heterogeneous with regard to professional rank. One observation should be made in passing this point, though we will return to it later. Very few of the highest ranking scientists in each of these sectors chose to use the services personally, but many of them encouraged their (subordinate) colleagues to do so. These users then filtered the information which they received from the services to their higher ranking (superordinate) colleagues.

The majority of our user group had Ph. D. degrees (Table 2).

The group was generally professionally young as indicated by the number of years beyond their bachelor's degrees (Table 3).

When categorized according to their fields of chemistry, we noted that their distribution is not significantly different from the national distribution (Table 4), with the possible exception of the overrepresentation of physical chemistry and the underrepresentation of biochemistry [1].

If we disregard graduate students (because of their professional inexperience), the user population shows a relatively high degree of professional involvment. The users belong to an average of four professional organizations per person, and have attended on the average of three open-national or closed-invitational professional meetings within the past two-year period. They are quite involved in information transfer as evidenced by an average of almost three speaking engagements outside of their home organization

TABLE 2

Educational Level

	Level	Percent (approx.)
1.	No bachelor's degree	0
2.	Bachelor's degree	26
3.	Master's degree	19
4.	Ph. D. degree	55
		100

TABLE 3

Professional Age

Years beyond bachelor's degree	Percent (approx.)
1-2	19
3-4	26
5-6	21
7-8	10
9-10	2
11-12	7
13-14	2
15-16	7
16 and over	7
	100

TABLE 4

Fields of Chemistry

Field	National (1)	Percent (approx.)
Analytical	11.8	9
Inorganic	5.5	5
Organic	36.7	41
Physical	11.9	30
Biochemistry	12.4	4
Chemical engineering	Not reported	5
Related chemical specialties	16.2 (other)	6
		100

per person over the past two-year period. On the average, they have presented two papers at professional meetings in the past two years. Finally,
many of the users serve as referees for scientific periodicals in their field
of interest. In all, we have a picture of a group of scientists highly involved
in their professions and the attendant processes of information transfer.

In order to estimate how much time our users devoted to keeping
aware of current developments in the literature, we asked them to estimate
how many hours per week they devoted to such activities. Although these
estimates are subjective, we take them at face value. We were struck by

the very serious spirit of the users when discussing their information problems, and the exactness of the manner by which they divide up their preci-
ous professional time. Therefore, these estimates seem to have a greater
intuitive validity than one might obtain perhaps for estimates of the time
spent in personal hygiene. When dividing the population into various subgroups—academic, industrial, etc.—we found no significant differences in
the amount of time that users devoted to current-awareness activities.
Therefore, the responses for the total population will be presented. Table
5 shows the users' estimates.

We then asked the users to divide this estimate into searching time
(time spent looking through indexes, titles, browsing articles, etc.) and
actual reading time (time spent reading fairly concentrated, selected works
from beginning to end) (Tables 6 and 7).

TABLE 5

Hours/Week Current Awareness

Hours	Percent (approx.)
0	1
1-5	32
6-10	43
11-15	11
16-20	12
21-25	1
26-30	1
31 or more	0
	100

TABLE 6

Searching Time/Week

Hours	Percent (approx.)
0	10
1-4	76
5-8	10
9-12	4
13 or more	0
	100

TABLE 7

Reading Time/Week

Hours	Percent (approx.)
0	1
1-4	35
5-8	39
9-12	16
13-16	8
17-20	2
21 or more	0
	100

We also wanted to know the number of journals to which our users personally subscribed. Surprisingly enough, a significant number of them subscribed to none (Table 8).

Most of our users had no previous experience with computer-based information retrieval services. Seventy-one percent claimed no experience, 21% had previously used one service, and 8% had used two services previously. Many of those with experience were users of the NASA information search system. Though their experience was limited, the users had a generally favorable attitude as to the usefulness of automatic information dissemination systems. Before submitting their first search profile, the users were asked to indicate their attitude (positive, none, or negative) on a continuum from +5 through 0 to -5. The results are given in Table 9.

The reasons which users gave for wanting to use the system were of four major types: 1) to save time, 2) to keep currently aware of developments in their field, 3) to cover more literature than is possible with conventional procedures, and 4) to help in our experiment and evaluate the usefulness of the PCIC. These categories are obviously not mutually exclusive or exhaustive.

We noticed from the demographic and professional statistics that we had a generally young experimental user population. Fifty-four percent are 12 years or less beyond their bachelor's degrees and a full 80% are at most 20 years beyond their bachelor's degrees. Many of the older professionals either chose not to subscribe to the services or, as mentioned earlier, suggested that their younger (subordinate) colleagues do so. Recalling that 55% of our users had their Ph.D.'s, it is worth noting that this is well above the 38.8% national figure for chemistry [1]. Most of these appear to be relatively young assistant or associate professors in academia

TABLE 8

User Journal Subscriptions

Number of subscriptions	Percent (approx.)
0	19
1	19
2	20
3	17
4	9
5	6
6	6
7	1
8	1
9 or more	4
	100

TABLE 9

Distribution of User Preconceived Attitudes

Attitude score		Percent (approx.)
-5 and -4		2
-3 and -2	Negative = 37%	11
-1 and 0		24
+1 and +2		26
+3 and +4	Positive = 58%	29
+5		3
Non-response		5
		100

and research scientists in industry. The point of this seems to be that the services are most attractive to young professionals who are still "on their way up."

III. OBTAINING USER FEEDBACK

We obtained feedback, through various procedures, from all subgroups of the user population which were using some or all of the information services offered by the PCIC. We used different methods to procure feedback in connection with each of the services involved: 1) Chemical Titles (CT), 2) Chemical Abstracts Condensates (CASCON), 3) the Automatic Subject Citation Alert (ASCA) System, and 4) Retrospective Searching of CASCON (Retro).

A. Chemical Titles

The CT service was the first to be offered by the PCIC. At this early stage of the project, we decided not to try to elicit regularized feedback due to inferior system reliability. We asked only for periodic relevance judgments from the 48 users, and concentrated on some unobtrusive indicators of system use. These are discussed later with the findings.

B. CASCON

The beginning of the CASCON service also initiated a regular and comprehensive system of procuring feedback from our users. The users completed entry questionnaires before beginning use of the service. They then received regular questionnaires at five- to six-week intervals which were intended to monitor their use of the system, provide a mechanism for users to complain about problems and make suggestions for system improvements, and to bring to our notice any changes in the users' information handling behavior as a result of using the service. To complement and supplement these questionnaires, we wanted to obtain relevant feedback about the use of specific alerts. In other words, we wanted to trace the life history of an alert. Until we did this, we only had rough aggregate statistics on relevance. We felt that there was a more interesting story to be told about what is done with an alert once it is received. In order to do this, we devised a sampling feedback procedure which could provide such information. We used feedback response cards upon which users answered questions about the use of specific alerts which were selected at random.

C. ASCA

To obtain feedback from the users of the ASCA service, we used forms equivalent to the CASCON questionnaires. Again, the users filled out an entry questionnaire and received follow-up questionnaires at five- to six-week intervals. In addition to this, a subgroup of users using both the ASCA and CASCON services were asked to complete a special questionnaire comparing the two services.

D. Retrospective Searches (Retro)

Feedback from Retro users was obtained through two methods. Users were first asked to fill out an entry questionnaire describing their needs for retrospective literature searching and the intended use of the output. After receiving their retrospective output, the users were asked by telephone interview to answer questions about its relevance and usefulness. The telephone interview is a useful technique for small groups of users where a high response rate is necessary.

IV. THE METHODS USED FOR EVALUATIVE RESEARCH

In general, we were interested in the mechanisms by which scientists procure and use information in order to satisfy their occupational information needs. We also wanted to discover how a particular information system was affecting, or at least related to, these mechanisms. The research environment, like any environment, affects the behavior of the people who live in it. We were interested in aiding their intellectual behavior by providing information services in the research environment. Therefore, a firm understanding of the research environment of the scientist was needed in order to place the information system in the proper perspective and context.

Each of the techniques we used will be discussed, pointing out their advantages and disadvantages and our successes and failures with them. In the following discussion it is well to remember that when a behavioral scientist does evaluative research on information systems, he is really testing the behavioral assumptions that are built into every information system by its designers.

Certain constraints should also be kept in mind for the ensuing discussion. When we talk of the value, effectiveness, or usefulness of an information service, we must be aware of the fact that we are dealing with a latent period of unknown duration. The true value of a service may not be determinable for months or years. One must be careful of the conclusions he draws from his work. Abstract statements can be drawn which have wide application for general information system design and implementation, but offer few specific directives for action in a particular system configuration. On the other hand, direct actions which are successful in a local situation may not be applicable elsewhere or generally. We have tried to avoid either extreme in our research. One final caution should be pointed out. No single data-gathering technique is satisfactory for all purposes. The evaluative researcher should use several methods to exploit his advantages, minimize his disadvantages, and cross-check his results.

A final note should be added before discussing each of the data-gathering techniques we have used. There are various vantage points from which to study user behavior: observe the user directly, observe the user through an agent, ask for a user's impressions of his own behavior, or simply question the user.

For each of our techniques, a certain amount of cooperation was necessary on the part of the user. The degree of their cooperation will determine what we refer to as the response rate. The response rate is simply the percentage of individuals of the total experimental population who provide a satisfactory response to the varied stimuli which are directed to them by the investigator, i.e., interviews, questionnaires, observations, etc. Enlisting the proper cooperation of experimental subjects can sometimes be a problem and many studies of user groups in the past have suffered from low response rates. It is often very helpful to offer some inducement to the subjects to encourage their cooperation. At the PCIC, subjects were given all available services on a free basis in return for their cooperation in responding to our stimuli. Those subjects who did not cooperate were then phased out of the free experimental group. In general, this strategy worked with great success and the response rates to all stimuli were consistently high. This strategy might be criticized for artificially "pumping up" the response rate and creating a bias reflecting free system use. This may be true, but there is also a definite bias associated with low response rates. Since we were in the position of trading-off one form of bias for another, we chose the one with the greater utility. Another factor which seemed to contribute, if only indirectly, to the high response rate was the continuing effort by the PCIC staff to establish and maintain good rapport with the experimental users. This was done by presenting seminars where experienced information analysts helped the users construct and modify their search profiles, and by making certain that communication channels were always open to the user in the future. In this way, the user was not alienated from the system.

A. The Unstructured Interview

We have used different types of interviews in different situations with considerable success. The unstructured face-to-face interview is a valuable tool for pretesting certain questions and eventually developing a structured interview schedule. This was done during the early stages of our project through three pilot phases, out of which came the structured interview. The questions used in the unstructured interview were open-ended and probing. The interviewer encouraged discussion, a technique which is often very valuable for suggesting new questions. It gives the investigator a broad general picture of the research domain from which fresh test hypotheses can be generated. At the early stages of our project, few of us had any past experience with information system design or evaluation, and as behavioral scientists we knew very little about the scientific subculture of the research chemist. The unstructured interview was a very helpful and instructive way of learning about these matters and provided a set of guidelines for our future work.

The unstructured interview does have its drawbacks, because it is very difficult to analyze in any firm quantitative way; and any results are, at best, very general. It is always best, if possible, to tape record any interview, and this is particularly important with the unstructured type. There is often so much discussion that the interviewer cannot record all of the important information. The tape recorder creates a more relaxed atmosphere and nothing is lost. Thus, the unstructured interview is chiefly a tool for developing strategies for the structured inverview.

B. The Structured Interview

The structured interview schedule was our most valuable tool for procuring detailed data about the research and information environments of the chemists in our user group. Structured interviews are often called "depth" interviews because they enable the investigation to probe deeply into relatively well defined ares. They are quite good for analytical purposes, since the respondent's answers can be placed into quantifiable coding categories which have been pretested and shown to be valid during the pilot phases when the final interview schedule is developed. The tape recorder is again valuable since it saves time and creates an atmosphere which is more oriented to discussion and less to answering questions. Time is the major drawback with the interview, structured or unstructured. Our original unstructured interviews were two and one-half hours in length, and the respondents showed considerable fatigue and sometimes hostility after about one and one-half hours. The final structured interview was no longer than an hour in length, and there was no apparent fatigue or hostility. One hour seems to be about the maximum safe duration for a successful face-to-face interview. If kept to a reasonable time schedule, the interview gives a wealth of detailed information, and a comparatively high response rate (near 100%) when compared with other forms of data collection. In addition, the interactive exchange between the interviewer and interviewee is valuable because it provides opportunities for clarifying questions for the respondent and enables the interviewer to gauge the respondent's mood and probe when necessary for more detail.

C. The Telephone Interview

Another type of interview which we used for a special purposes was the telephone interview. This technique was employed with the Retro users for several reasons. It is easy to administer, in that the interviewer does not have to be mobile and go to the respondent's location. It was short and to the point and thus well suited for the telephone. We felt that telephone interviews should be much shorter in length than the face-to-face interview, since the call was not prearranged as was the face-to-face situation. Of course, if the respondents did not have the time to talk at the first calling, arrangements were made to return the call. Most important were the facts that 1) the project's experimental effort was drawing to a close, and in the time remaining, the telephone interviews could be quickly completed; and 2) that the group of Retro users was relatively small, and we needed the higher response rate which the telephone interview could insure in contrast to a mailed questionnaire.

D. Questionnaires

Questionnaires were, at one time or another, filled out by all of the experimental users. This approach provides a very good means of reaching a geographically scattered population while still being fairly inexpensive. It is effective when asking individuals to report "typical" behaviors but not when asking them to relate particular occasions with any detail. It is useful for getting answers to special questions about which the respondents are surely familiar and have easily recallable answers. Questionnaires do not provide the investigator with any knowledge of the respondent's general receptivity to the questions asked and do not afford any opportunity to clarify questions or to make certain that the respondent fully understands the question and will respond properly. As a result, there is no chance to probe more deeply for details and the subject is forced to respond within a limited range of fixed response categories.

Every item on a questionnaire should be checked against the following criteria. This is true for any stimuli such as interviews or observation checklists, but it is particularly crucial for the questionnaire for the above-stated reasons.

1. What information is sought?
2. What is the most appropriate sequence of the topics to be covered?
3. Does the respondent have the necessary information to answer the question?
4. Is the question's content or wording biased?
5. Can the question be understood or does it contain unclear phraseology?
6. Does the question provide for all of the possible alternatives of response? (checkmark, continuum, write-in)
7. Is the form of response best suited for a proper answer?

8. Does the content of previous questions bias the response to the present question ?

9. Is each question absolutely necessary or can the information be obtained elsewhere or at another time ?

If possible, the questionnaire should be pretested on a pilot group of subjects before general distribution. (All of our questionnaires were first pretested on the library committee of the Chemistry Department of the University of Pittsburgh).

Finally, one of the most serious disadvantages of the mailed questionnaire is the relatively low response rate. It leaves the respondent with the entire responsiblity of completing the questionnaire and returning it by mail. Again, certain inducements for cooperation are helpful, but making the questionnaire as simple and as easy to process as possible is always best. If the response rate is still low, more effort may be required on the part of the investigator. A short telephone interview would be the next best alternative.

We have mentioned that the users of CASCON received longitudinal questionnaires at five- to six-week intervals for nearly a year. The intent was to enable us to monitor the user's response to the service and to note the changes that would take place in his work, in his handling of information problems, in his productivity, in his attitude toward the system, etc. No significant changes were found. The responses on repeated questions over the year were amazingly consistent with previous responses. Slight changes in the hypothesized direction were evident (i.e., saving more time, feeling better informed, gaining greater confidence in the system), but they were not significant—just encouraging and intuitively satisfying. We conclude that this innovation in the information environment of the research chemist either did not affect his behavior or that it has a latency longer than one year. Perhaps more time is needed before the full effect of these services becomes apparent.

E. Unobtrusive Techniques

Unobtrusive data-gathering techniques are often effective when it is not possible or not desirable to have direct contact with the user. The intent is to obtain and analyze the records or results (products) of the user's application of the sytem. One can use library or system-use records, perform request studies of inquiries at information centers, or perform relevancy studies by tracing citations from alerts. Unobtrusive techniques are especially good when used to supply supplementary information in combination with other methods and to compare the user's perception of what he did with the records of his actions. Often, records such as these provide valuable data which either goes unanalyzed or underanalyzed.

F. Alert Response Cards

   We wanted to know more about the fate of the CASCON alerts received
by the users than would be gained from crude relevance assessments;
namely, what actually happened to the alert, did it lead the user to a document, and how was the information in the document used? Conversely, we
wanted to determine why individual alerts were rejected: were they just
not relevant, were they from an obscure journal, or were there no trans-
lations available from the foreign language? We had to devise a strategy
that could answer these questions and still not be too demanding upon the
user. Each user received his alerts on double IBM cards connected with a
center perforation. The two halves of the card were identical and contained all of the bibliographic information necessary for the user to retrieve
the document. If the article was of any interest to the user, he was instructed to keep the left half for his own files and return the right half to
us; otherwise, he returned the whole card. The ratio of half-to-whole
cards returned was thereby the user's relevance percentage; and since
the cards were sequentially numbered and punched, we were able to keep
track of every alert received by each user. An example of the card is
shown in Fig. 1.

   After the users were familiar with the above precedure, an additional
task was required of them. Each user had 10% of his alerts randomly
stamped "PLEASE RESPOND" up to a maximum of 10 alerts per search,
i.e., if a user received 60 alerts, six of them would be marked for response. For these alerts, the users followed the same whole-half procedures, but in addition they responded to multiple-choice questions on a

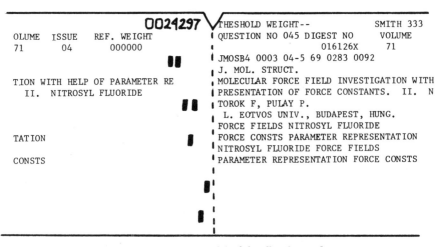

FIG. 1. Example of feedback card

sheet accompanying the output. Appropriate places were marked on the reverse side of the card (see Fig. 2) regarding the accompanying alert. Since the alerts for every one of the users were randomly sampled, the results of the study provided a representative picture of the use of the alerts for all users.

We were able to 1) determine whether or not the users had previously seen the documents to which they had been alerted and 2) determine their degree of usefulness far beyond simple relevance. Several reasons for the rejection of alerts were determined beyond nonrelevance. We were able to ascertain whether the subjects felt that the information they had received was "basic" or "applied" and at which stage of their research it was most useful. Total average relevance of the CASCON alerts for the period studied (Vol. 71 of Chemical Abstracts) was 44%, but this and other findings will be discussed later.

V. STATISTICAL TECHNIQUES

We have used percentage statistics for purely descriptive purposes and in the creation of certain categories. These involve the reporting of percentages and describing how subjects distribute according to certain single variables. Examples of these are shown in the early part of this chapter, e.g., educational level, professional age. For purposes of statistical association and inference, statistics which investigate the relatedness of two variables were used. Chi-square contingency table analyses were used successfully in determining the degree of association

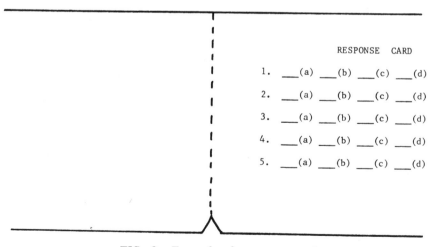

FIG. 2. Example of response card

	Industrial	Academic	
User	35	18	53
Nonuser	15	32	47
	50	50	100 = Total

FIG. 3. Example of contingency table.

between sets of variables [2a]. For example, we might have a situation as shown in Fig. 3. As can be seen, the table is really the joint-frequency distribution of two variables. Chi-square analysis investigates the degree of co-occurrence of the two variables to determine their statistical association. The contingency chi-square test shows that two variables are associated.

A stronger and more precise test of the statistical association of two variables is achieved by using correlational techniques. According to the Pearson produce-moment correlation technique, two variables are unrelated if their coefficient of correlation is zero [2c]. When two variables are maximally related (the two variables increase at the same rate), their coefficient of correlation is +1.0. When two variables are inversely maximally related (one increases while the other variable decreases at exactly the same rate), their coefficient of correlation is -1.0. Thus the correlation of two variables ranges between +1.0 and -1.0 and indicates the similarity of two sets of scores on the two variables. If two variables are positively correlated, subjects with high scores on one variable would also tend to have high scores on the other. If the variables in question are negatively correlated, subjects with high scores on one variable would tend to have low scores on the other.

One problem which continually confounds behavioral scientists and was encountered in our evaluative research effort is the relationship between correlation and causation. When two variables are highly correlated, this does not imply that one necessarily "causes" the other. Their correlation may be accidental and their association due to their relation with some third variable which may be "causing" them both. Often, the third variable is not known. This situation necessitates the use of a more advanced form of correlation known as partial-correlation, where one can determine the degree of association between two variables while holding the effects of a third variable on the first two constant.

Multivariate statistical analysis is a valuable tool for a variety of purposes and allows the simultaneous analysis of data derived from measurements on three or more variables. When we have N subjects, measured on each of m variables, each of the N subjects can be character- ized by a point reflecting their scores on the m variables projected into m-dimensional space. During the early, exploratory stages of our work, multivariate techniques such as principal components factor analysis were useful in uncovering underlying relationships which previously were not felt to exist between the variables. Thus, multivariate techniques can test old hypotheses and suggest new hypotheses to guide future research. Factor analysis is often useful for defining meaningful dimensions in a research domain. This is done by creating an index or scale. Several variables are combined according to their high correlations, in order to suggest a new pattern of their own and thus actually create a new variable on a higher level of abstraction. For instance, in studying urban problems one might combine income, education, occupation, ethnicity, race, and place of residence to produce an index which forms a new variable called "socioeconomic status." In our own work, we combined the following set of variables to arrive at an index of professional involvement for a sub- group of experimental users. This index was then related to variables of information processing.

1. The number of journals for which the subject served as editor.
2. The number of journals for which the subject served as referee.
3. The number of memberships in professional organizations.
4. The number of professional meetings attended in the past two years.
5. The number of papers delivered at professional meetings in the past two years.
6. The number of times the subject actively participated in a profes- sional meeting in the past two years above and beyond paper pres- entation, i.e., chairman, discussant, organizer, keynote speaker.
7. The number of times the subject has given invited lectures or colloquia at a college or university other than the one(s) he attended.

The statistical techniques used most often were principal components factor analysis, multivariate analysis of variance, and multiple discrim- inant analysis. All of these are important because current research con- tinually suggests that the scientist's information needs and uses are not explained by a single attribute, but by a combination of interrelated vari- ables. The usefulness of these techniques for multidimensional analysis will be described briefly. Some of the empirical uses of factor analysis have already been mentioned, but several more exist. In general, all of these are concerned with the analysis of the intercorrelations among a set of variables. Principal components factor analysis is one of a variety of factor analytic techniques which enables the researcher to identify the orthogonal (independent) dimensions that are necessary to explain the variance in scores on a group of variables. Thus, the original dimensionality (the number of original variables) can be dramatically reduced to a few principal components which account for most of the original variance.

Multivariate analysis of variance is based on a generalization of the F test which is used in univariate analysis of variance. One seeks to determine whether two or more groups are significantly different as reflected by their members' scores on a set of variables. Tests for equal group variances are first made; and if that assumption holds, tests for significant differences between group centroids are made. (Centroids represent the groups' central most position in the multidimensional variable space, and are analogous to a body's center of gravity.) If the groups are significantly different, it is interesting to know the magnitude of their differences and how much each of the original variables are contributing to discriminating between them. Multiple discriminant analysis can be used to answer these questions. Discriminant analysis allows the researcher to use a linear function of the scores on the original variables to construct a line (the discriminant function) which best discriminates between the experimental groups. One can then determine the contribution of each of the original variables to the discriminant function, which thus indicates the ability of each of the variables to discriminate between the groups. The position of the group centroids on the derived function can then be used to indicate the magnitude of the group differences.

VI.   CURRENT-AWARENESS LITERATURE SEARCHING

The need for current-awareness information is probably most crucial to the working research scientist, with the need for retrospective literature searches a close second. However, the priority of these needs also depends upon the individual user's style of research. Many current-awareness users have been working in a field of chemistry for several years (possibly their entire professional lifetime). These individuals have little need for retrospective searches (except under special circumstances discussed in Chapter 7) because they are already aware of the significant developments of their field. They probably have a file or bibliography as comprehensive as any machine search would be in their particular fields of interest. These individuals may find that a retrospective literature search is valuable if it is broad enough to include areas of chemistry that are closely related to their principal interest, but this often requires a broad strategy, which exposes them to the risk of being bombarded with many irrelevant and uninteresting alerts. Regardless of their need for retrospective searches, scientists with a continuing interest in a field are compelled to keep up with the latest developments in their area of interest, and are in great need of current-awareness information. Competition in their field is often high, and they need to be kept aware of priorities of discovery, patents, developments in technique and apparatus, and the new research findings of their colleagues and/or competitors.

A. Chemical Titles Study: The Use of Unobtrusive Indicators

The records of the use of Chemical Titles searches provided the data for our first analysis. We wanted to determine what they might indicate about the use of this service by various subgroups of users. The records included such items as the number of terms per user profile, use of truncation modes, number of alerts received, and cooperation in providing feedback. Analysis of the feedback data showed that early CT users (those who took advantage of the service as soon as it was available) had a significantly higher number of search terms per profile than the late users (those who came after the service had been operating for one month). The early group consistently used left-truncation, reflecting more general search strategies, whereas none of the late group used it. The early group got more from the service, in a quantitative sense, averaging 38 alerts per search as compared to 14 for the late group. Finally, they consistently provided more frequent feedback (in terms of relevance judgments) than did the late group.

Those who used the service at the first opportunity seemed to be more aggressive information seekers, and the more these users asked of the service (more parameters, more search questions, more search terms) and the more they got from the service (more alerts), the more cooperative they were in giving the feedback about the service's performance. These users had broad information needs and broader areas about which they wished to be kept currently aware, and the system seemed to fulfill the need for breadth of coverage. Those users, on the other hand, who did not ask much of the system and did not get much from it tended to feel alienated from the system and felt that it did not or could not satisfy their needs.

This analysis, as was the case with many to follow it, first pointed up the great variability in styles of system use, i.e., low number of search terms vs a high number of search terms, complex vs simple strategies, use of truncation vs nonuse of truncation. This and other studies helped confirm what many of us had felt intuitively— great user variability is something which must be taken into account in constructing and implementing a system to serve the information needs and uses of research scientists. There should be a great deal of flexibility in such a system to allow for the large amount of individual differences found in the population it hopes to serve. Individual user differences and a good deal of idiosyncratic user behavior should be taken into account in order to provide optimal service.

B.   Chemical Abstracts Condensates: User Feedback

Several of the findings concerning reactions to CASCON have already been presented and discussed, but more detailed information about user feedback should be presented to gain a full picture of the use of the service. The overall response rate to our questions was approximately 60% over the eleven-month period of the experiment, during which the number of users responding varied from 53 at the beginning to 103 at the end.   Therefore, percentage figures will be reported to indicate the distribution of feedback response to the questions during this period.

We found that 58% of the CASCON users (response N = 103) were using the service for themselves and their own interests, but that 42% were using the system on behalf of a group and were passing along alerts to other members.   If a group is working in a well-defined research area, it is often useful to have a broad group profile to maintain current awareness of the area.   (This should be important to vendors of chemical information, since it means that they should allow for and even encourage group profiles.)   We found that information sharing and passing on information to members of one's research group is common.   Our users felt that such information interchange with colleagues was a beneficial element of the research scientist's information environment.   A similar study at the E. I. DuPont Company showed that 76% of the CASCON users reported passing on one or more of their alerts to other scientists within the firm [4].

We were constantly aware of the problem of overburdening our experimental users by requiring them to devote too much time to the mechanics of using our services, e. g., profile modification, general communications, or answering our questionnaires.   When we asked our users to estimate the amount of time per month that they actually devoted to using the system, they (response N = 66) responded as shown in Table 10.

TABLE 10

Amount of System Use

Hours/month	Percent
0-1	42
1-2	23
2-3	12
3-4	8
4-5	7
5-6	3
6 or more (to 9)	5
	100

Thus, 77% indicated that they devoted up to three hours per month in actual system use. Users could also be overburdened by receiving more alerts than they could fruitfully evaluate in a two week period. When our users were polled on this question, their response was (N = 59) as shown in Table 11. Users receiving more alerts than they could fruitfully evaluate often found that making their profiles less broad would cut down on the incidence of "noise" alerts in their output. However, many users were willing to put up with a considerable amount of noise in order to find useful alerts in their output. We asked our users (response N = 93) to estimate the percentage of nonuseful alerts (of the total amount received at each two-week output) that they were willing to tolerate, i.e., what percentage they were willing to "weed-out" (Table 12). A factor which some users reported as being annoying was that CASCON sometimes alerted them to documents about which they were already aware, i.e., redundant information. Some users, on the other hand, counted this as a "plus," in that it was often reassuring to have different techniques of current awareness intersect on the same document. Some said that this increased their confidence in the recall of both intersecting services. Nevertheless, we asked our users to estimate the percentage of alerts of which they had already been made aware from some other source, e.g., manual literature searching, ASCA, Chemical Titles. We found (response N = 84) the results shown in Table 13.

The alert response cards provided additional information about the use of the CASCON service. One hundred eighty-one users, two-thirds from academic institutions, cooperated in this study by returning some or all of their alert response cards. They returned the alert answer card attached to a random 10% sample of their alerts, and overall we had approximately a 40% response rate. Generally, most users were not previously aware of an article to which they had been alerted (86% of the articles fell in this

TABLE 11

User Evaluation of Alerts Received

Alerts fruitfully evaluated	Percent
0–25	39
26–50	22
51–75	8
76–100	15
101–150	7
151–200	5
Over 200	4
	100

TABLE 12

User Toleration of Nonuseful Alerts

Percent not useful	Percent response
0-19	2
20-39	14
40-59	23
60-79	31
80-100	30
	100

TABLE 13

Redundant Alerts

Percent redundant	Percent response
0-19	34
20-39	26
40-59	18
60-79	11
80-100	11
	100

category). Those who were had usually come across them in their regular
reading. The relevancy of 83% of the alerts was decided on the basis of the
bibliographic information given on the alert card. Only 3% of the alerts led
to an abstract actually being read, and in only 4% of the cases was the
source article read before deciding on relevancy. Over one-third of the
alerts were useless, and they were rejected usually because the information
contained on the alert indicated that the document had a low potential and
did not appear to be worth the effort required to follow it up.

C. ASCA

The ASCA service was offered to scientists at the University of Pittsburgh, Harvard University, Carnegie-Mellon University, Mellon Institute,
and the Koppers Company Research Laboratory on an experimental, non-
fee-paying basis. There were 64 experimental users of ASCA, and 54

of these were simultaneously using one of the other two current-awareness services, CASCON or Chemical Titles. Although the ASCA and CASCON services differ to some degree, there seemed to be no significant behavioral differences between those who used ASCA and CASCON. The ASCA users were most pleased with the results of citation searching and with the currentness of the ASCA alerts, but some felt that its literature coverage was too narrow.

D. Comparison of CASCON and ASCA Services

The 37 scientists who were using both the CASCON and ASCA services were asked to compare them in terms of specified criteria. Eighteen had used ASCA prior to CASCON, and nine has used CASCON prior to ASCA. Nineteen of the 37 felt that it was beneficial to use both services and gave three main reasons.

1. Improved coverage: The services were complementary, uniting the strong features of the two (CASCON'S completeness and general coverage and ASCA's citation searching).

2. Service intersection: Important papers could often be identified through redundant alerts from both services.

3. Complementary coverage: ASCA was used to retrieve references for a specific research problem, while CASCON was used to uncover secondary references from a more general research interest area.

Generally, the respondents to the comparative questionnaire (68% response rate) found ASCA slightly more useful for their current research interests. ASCA was considered much easier to use than CASCON. CASCON was felt to have considerably broader scope and breadth of literature coverage, though it often produced more useless information than did ASCA.

VII. THE PREDOMINANCE OF THE SCIENTIFIC JOURNAL

For the majority of chemists, the scientific journal is considered to be irreplaceable as the point of access to the chemical literature. Structured interviews were used to determine how various media of scientific communications fulfilled specified functions of information processing. The respondents were given a list of media (see below) and asked to indicate which medium best fulfilled each of the following functions: 1) to give general awareness of the current state of their field of chemistry (AWARE), 2) to find out who is working in what area or on what problems (WHO), 3) to serve as a source of specific ideas for work in progress (PROG), 4) to serve as a source of specific ideas for new work (NEW), 5) to inform others of their research activities (INFOR), and 6) to provide general browsing and stimulation (BROW).

In every case, the scientific journal (recent journal articles) was selected as the most important medium for each of those functions. We found (response N = 42) the results given in Table 14.

Media of Scientific Communication

1. Recent journal articles.
2. Reprints of more dated articles.
3. Manuscripts, drafts, or preprinted material.
4. Technical reports distributed within own institution.
5. Technical reports distributed by other than own institution.
6. Telephone conversations.
7. Face-to-face discussions with persons working at your own institution.
8. Face-to-face discussions with persons not currently working at your own institution (e. g., at scientific meeting, etc.).
9. Oral presentations made at scientific meetings, conferences, or seminars.
10. Copies of oral presentations, including lecture notes and conference proceedings, papers given at meetings, etc.
11. Private correspondence.
12. Review articles.
13. Reference books.
14. Text books.
15. Other books.
16. Dissertation Abstracts.
17. Chemical Abstracts.
18. Chemical Titles (published version).

TABLE 14

Functions of Information Processing Fulfilled
by Scientific Communications

Function	Percent choice of recent journal articles
1. AWARE	55
2. WHO	45
3. PROG	52
4. NEW	41
5. INFOR	43
6. BROW	50

19. Technical clearinghouse publications of government agencies.
20. Technical advertisement.
21. Science fiction.
22. Patent literature.

In every case, those who chose a medium other than the recent journal article, chose one of several others. No other single medium even approached the journal article in importance. These results should undoubtedly be taken into consideration by anyone who would consider replacing the journal as a component in a scientific transfer system.

VIII. FACTORS INVOLVING THE USER'S ORGANIZATIONAL AFFILIATION

Numerous factors have to do with the particular organizational setting from which a user comes. Several users from academic institutions discontinued the services when told that they would have to pay after being phased out of the experimental user group. Many explained that they felt that the services were quite valuable, but that during the depressed state of scientific funding (1969-70) they were primarily concerned with just keeping their experimental efforts alive and all research funds had to be devoted to that end. The cost of the services at the University of Pittsburgh and Carnegie-Mellon University was paid by the library system, but those at other academic institutions did not have this advantage.

Industrial users have some special problems, one of which is the protection of proprietary rights. Many industrial users, when they were beginning to use the services, were greatly concerned about the security of their search profiles and output. Some were worried that a competitor might be able to determine what problems they were working on, either from seeing their profile or their output. Accordingly, a numerical identification system was used to preserve the industrial user's anonymity. In addition, the industrial users from the larger concerns had their system interface handled by an information officer from their own research center. This worked well since the information officers were either librarians or people involved with the company's internal information services. Often the services, particularly during the experimental period, required a good deal of monitoring and system interface on the part of the user. We found that the information officers in the large industrial concerns took much of this burden from their users.

Of all the companies which subscribed to the services during our free, non-fee-paying experimental phase, the only ones which chose to discontinue the services on a fee paying basis were the eight smaller industrial companies. The reasons given for discontinuing the services could be summarized as follows: many of the smaller companies had very limited

product lines, small research staffs, and a relatively narrow technological
base. As a result, they felt that they could not use all of the information
which they received, because they did not have access to libraries to procure the documents and they did not have the manpower to act upon the information. For the users who dropped out, the breadth of coverage which
was provided by the services was not worth the time required to monitor
the system and search for the documents which might have been valuable
to them. On the other hand, the users from larger industrial companies
with many product lines and a relatively large technological base, were
quite happy with the services. They had libraries which were easily accessible and librarians who monitored a large amount of the system interface for them. The librarians were helpful in procuring documents and
much of the "leg work" was removed. The research staffs were often
large enough so that alerts could be passed on to other interested colleagues.
In this way there was a good possibility that the alert could be acted upon
by someone. If the company's line was broad, the probability that an alert
(and therefore the service) would be useful to the company was thus increased. The current-awareness services, therefore, seemed to be best
received by users from large industrial concerns.

On the other hand, retrospective literature searching seems to be desired by researchers of both small and large industrial companies. The
reasons seem to be similar for both. They are often called upon by the
management of their firms to become involved in areas with which they
have previously had little or no research experience because new resources
have been found, product improvements or changes are necessary, or a
new area of research has come into vogue, e.g., antipollution, ecological
research, product-recycling. The industrial researcher is often told to
stop what he is presently engaged in and to begin work on some area which
is either entirely unrelated or at best only tangentially related to his previous interests. In such situations, the value of a quick retrospective
search of the literature of the new interest area is very high.

IX. BASIC AND APPLIED RESEARCH SCIENTISTS

Several previous studies of the information processing activities of
scientists have relied on an a priori classification typology centering on
the so called "dichotomy" of basic and applied research. The subjects
were assigned to one or the other types according to their organizational
affiliation. This dichotomy might provide a good approximation to the
true research interests of the subjects, but we find that there is applied
research in academic environments, and basic research being conducted
in industrial and governmental research laboratories. We needed an indicator of each scientist's research interests which would provide a more

accurate way of classifying him as basic or applied than relying solely on his affiliation. The problem had to be conceptualized in terms of a continuum from most basic to most applied. We felt that since the time research scientists devote to the current-awareness literature of their field of interest was so highly valued, this might provide a good indicator. We had a panel of expert judges rank the most frequently used journals in chemistry on a continuum (from five to one) from basic to applied. We then assigned a score to each journal from the average of the scores given by the panel. We developed a list of journals which were ranked on the continuum from most basic to most applied. The expert panel of 68 CASCON users provided ratings for 115 journals from chemistry and closely related areas. Each journal received a continuum score in the possible range from five (most basic) to one (most applied). The actual range of scores was 4.50 for the journal whose content was most basic to 1.46 for the most applied journal.

Using the structured interview schedule, each of 70 experimental subjects was given the list of journals (without knowing their continuum scores) and asked to indicate which of the journals on the list he habitually scanned, read, or used in any way for his professional work. Each individual received a score which located him on the basic to applied continuum. Each scientist's score was computed as the average of the scores of the journals which he mentioned. In this way, we had a precise indication of the interests of each of the subjects as reflected by their relative ranks on the continuum. The actual subject ranks ranged from 4.21 (the scientist with the most basic research interests) to 2.28 (the scientist with the most applied research interests). It was found through statistical tests that these scores were normally distributed. This work had been a major step toward breaking the old sterotypes of "basic" and "applied" scientific researchers.

The group of 70 subjects was then divided into three subgroups by converting each subject's continuum score to a standard or Z-score. The Basic group (n = 25) was composed of all subjects, $Z \leq -.5$, the Mixed group (n = 28) was composed of all subjects, $.5 < Z > -.5$, and the Applied group (n = 17) was composed of all subjects with Z-scores, $Z \geq .5$. Using multivariate analysis of variance, the three groups were found to be significantly different (beyond the .05 confidence level) with regard to the seven variables of professional involvement (see section V, i.e., the Basic group was most professionally involved and the Applied group was least professionally involved. The Basic and Applied groups, when considered alone, were significantly different beyond the .01 confidence level on the same variables.

Disregarding the Mixed group for the time being to simplify the analysis, let us compare certain aspects of information processing of the remaining Basic and Applied groups. The subjects were given four hypothetical situations in which they were asked to indicate which of eight methods of information gathering they would use first, second, and so on. The order in which the eight methods were presented to the subjects was changed in each of the hypothetical situations. The four hypothetical information procurement situations were the following.

1. You are working on a design for a procedure or experiment in a field in which you are familiar and you wish to know if similar work has been published recently or is currently being done by someone else.

2. You are preparing a proposal (involving approximately $60,000) for a new project either to the management of your organization or to an outside granting agency. You wish to substantiate the proposal with a thorough bibliography.

3. You want to gather information in order to write an article in your area of specialization for a review journal.

4. You are about to begin work on a research project in an area in which you have not done any previous research.

The eight methods of procurement were: 1) search your personal library, 2) search material within your organization's library, 3) use some other library outside of your organization, 4) consult a reference librarian, 5) telephone a knowledgeable person who may be of help, 6) visit a knowledgeable person within your organization, 7) visit a knowledgeable person outside of your organization, or 8) write a letter to request information from a knowledgeable person [3].

The first, second, and third choices for each subject were pooled so that there would be 75 choices (3 X (n=25) = 75) for the Basic group and 51 choices (3 X (n=17) = 51) for the Applied group and a total of 126 (51 + 75 = 126) choices. Notice that the first four methods of procurement on the list are nonpersonal, literature-oriented sources, while the second four methods are nonliterature, person-oriented sources of information. In order to take account of the relative frequencies with which the personal methods are used as compared to the literature-oriented methods for both the Basic and Applied groups, we construct contingency tables for each of the four situations, and use the chi-square test for the independence (nonrelation) of the Basic and Applied groups. Since we are dealing with three dependent responses (i.e., a person's second or third choice will depend on the ones preceding it), we must take one-third of the value of each chi-square which is generated. The results of the analysis are shown in Tables 15, 16, 17, and 18.

The value of chi-square is not significant, and therefore in the first hypothetical situation there is no significant difference in the choices of procurement methods between the Basic and Applied groups (Table 15).

This chi-square value is zero due to the fact that the expected cell frequencies were identical to those observed, indicating that there is no

difference between the two groups in the types of procurement methods used
in this hypothetical situation (Table 16).

TABLE 15

Procurement Situation One: Basic vs Applied

	Basic	Applied	
Methods 1-4	51	38	89
Methods 5-8	24	13	37
	75	51	126

TABLE 16

Procurement Situation Two: Basic vs Applied

	Basic	Applied	
Methods 1-4	59	41	100
Methods 5-8	16	10	26
	75	51	126

$1/3 \, X^2 = 0.0$ with 1 degree of freedom

This value of chi-square is significant beyond the .05 level. The Basic
group, in this situation, ·is significantly more dependent upon the nonpersonal,
literature-oriented methods of procurement (Table 17).

TABLE 17

Procurement Situation Three: Basic vs Applied

	Basic	Applied	
Methods 1-4	62	29	91
Methods 5-8	13	22	35
	75	51	126

$1/3 \, X^2 = 4.01$ with 1 degree of freedom

TABLE 18

Procurement Situation Four: Basic vs Applied

	Basic	Applied	
Methods 1-4	48	17	65
Methods 5-8	27	34	61
	75	51	126

$1/3 \, X^2 = 3.77$ with 1 degree of freedom

This value is significant at the .05 level. The Basic group is again more dependent on the nonpersonal, literature-oriented methods of procurement, while the Applied group is more dependent upon the personal, non-literature-oriented sources (Table 18). We can see that procurement behavior varies considerably with the information-seeking situation and that differences between the Basic and Applied groups were only evident in the last two situations.

It is interesting to note that in situations one and two both the Basic and Applied groups preferred the literature-oriented methods of procurement over the personal methods: Situation One = 89 to 37 and Situation Two = 100 to 26. In the final two situations there were significant differences between the Basic and Applied groups. The first two situations are undoubtedly more common than the latter two, and this may account for these differences, but it should also be pointed out that the latter two situations are ones in which a retrospective search of the literature might be warranted. The fact that the applied researchers avoided the literature-oriented sources suggests that they do not readily perceive an easy method of access to past literature. One could speculate that this indicates a greater need for automated retrospective literature searching among applied researchers than those in basic research. The basic researcher may feel more in contact with the literature in the latter two situations and not feel the need for personal assistance in procuring the needed information.

X. THE EXPERIENCES OF A RESEARCH GROUP: A CASE STUDY*

The main research group to be studied was located at the University of Pittsburgh and has its research interests in physical organic chemistry. A second research group located at Harvard University was also studied to determine if the same general patterns of service use emerged. The reactions of this group will only be mentioned if they strongly support or significantly differ from the reactions of the Pittsburgh group. The case study proved a very effective way of obtaining both detailed and personalized information about how the members of the research groups used selected information services. The responses from members of these groups are not necessarily representative of the responses of all academic chemists, but they are, at least, informative.

It was felt that the Pittsburgh group's current-awareness activities could be studied by encouraging them to use any one or all four current-awareness literature searching techniques. This group was composed of a research director, two postdoctoral fellows, and three graduate students working on their theses. Most of the 17 Harvard users were from one research group in organic chemistry. There were, in all, three faculty members (one full professor and two assistant professors), four postdoctoral research fellows, nine advanced graduate students, and a science librarian.

The members of the Pittsburgh group were provided with copies of the tables of contents of the journals which they felt were important for their work, and each checked the articles of interest. A second technique was employed, whereby the members of the group scanned the Physical Organic

---

*Much of what is to follow is abstracted from Ref. 5.

Chemistry and Thermochemistry sections of <u>Chemical Abstracts</u> and checked the articles of interest. The third technique used by both groups was the Automatic Subject Citation Alert (ASCA) Service, a computerized system searching about 600 chemical journals. The fourth technique for current awareness which was offered to members of both groups was <u>CA Conden-sates</u> (CASCON). This is a computerized service which searches the listings of all items in a single issue of <u>Chemical Abstracts</u>. Since this was the major service offered by PCIC, its use was studied in greatest depth.

For the Pittsburgh group, all of the articles checked by the first two techniques and all of the relevant alerts from the second two computerized techniques were divided by an expert into two categories: 1) group interest, i.e., those references of direct interest to the group research activity; and 2) general interest, i.e., those references of interest but not directly related to the group's activity. Certain of the findings of the use of the four current-awareness techniques are striking. Almost 18% of all of the total relevant references chosen by the group (including both categories) were from the <u>Journal of the American Chemical Society</u>; 6.4% were of <u>group interest</u> and 11.1% of <u>general interest</u>. In the five months of the study, more than 1465 references were involved, and of this total, 40% were group interest and 60% general interest.

The Pittsburgh group selected a total of 25 journals whose tables of contents they regularly scanned. Of these 25, nine were selected by four of the six group members, eight by three of the six group members, and the remaining eight were suggested by one or two of the group members. Of the total group-interest references found, 60.6% were from these 25 journals, and 73.6% of these were found by scanning the tables of contents. When members of the group found references of particular interest, they could request a copy of the first page of the article. Of the 177 requests for references of particular interest, 129 (72.9%) were from these 25 journals.

The Pittsburgh group found that scanning the two <u>Chemical Abstracts</u> sections was a successful current-awareness tool. Over a sixteen-week period, an average of one out of every five abstracts in the two sections was selected as being of interest to one of the members of the group. Seven percent of the group-related items and 18% of the general interest items were selected using this technique. Although these percent figures are low, the relevance percentage of these references was very high because the members of the group had the full abstract available before making their requests.

A very broad ASCA search profile containing nine author's names and 121 bibliographic citations was constructed for the Pittsburgh group. Fifty-one percent of the total retrieved items were judged as being relevant

by at least one member of the group, while total individual relevance figures ranged from 2% to 28%. It should be remembered that this is a broad
group profile and individual profiles have much higher relevance. Even
though ASCA covers 600 chemical journals, 72% of all its relevant alerts
were from the 25 journals selected by the group as the most important for
their interests. For some unknown reason, there seemed to be very little
interest in the ASCA service among the Harvard group. Only four short
profiles were submitted, and because of this they could not properly reflect
the potential usefulness of this service to the user.

Each member of the Pittsburgh and Harvard groups submitted an interest profile for the CASCON current-awareness search system. We found
that the members of the Pittsburgh group averaged one hit per five alerts.
As compared with the other techniques, CASCON sometimes presented a
timing problem for members of the group. CASCON is related directly to
the production of Chemical Abstracts and must therefore wait until the full
abstracts are put in the data base and made searchable, while ASCA and
CT rely only on authors and titles and are operative as soon as the journals
are published. The items retrieved with CASCON thus come at a time
significantly after the journal has arrived on the library shelf, i.e., 90%
came within 14 weeks after their corresponding journal had reached the
library. The significance of this delay factor varies greatly with each
user. Being alerted to an item at an early date may be helpful if it is
directly related to the user's ongoing research endeavor, because it could
save possible duplication of effort. Early alerting is also important in
cases where priority rights are at stake, i.e., the case of patents. On the
other hand, for a general awareness of the progress in a field, the delay
is usually not considered significant by the user. He is glad to trade off
the broader coverage afforded by CASCON for any discontent with its delay.
Regarding the delay question: of 58 users sampled (primarily from academic institutions), 31% felt the tolerable time lag was one month, 29%
said two to three months, and 22% four months or more (16% did not respond).

To compare the recall of the four current-awareness techniques used
by the Pittsburgh group, 500 references were selected at random from all
those retrieved. These were known to be in each of the four data bases
used in the experiment. Forty-three percent of them were found by scanning tables of contents, 28% from the use of the two Chemical Abstracts
sections, 35% from ASCA, and 67% from CASCON.

Of the Pittsburgh group's six members, the involvement with the chemical literature ranged from a high degree of interest on the part of one
postdoctoral fellow to a very narrow interest in his thesis topic expressed
by one graduate student. When the members of the group could delegate
the task of locating and retrieving copies of items of interest to other

persons, they were less selective in their relevancy judgments than when they had to retrieve the articles themselves. In the latter case, they were likely to do more screening to get the most important articles. ASCA was generally felt to be more valuable than CASCON, since it provided citation searching. There may have also been a latency factor contributing to the preference of the Pittsburgh group for ASCA, since they had been using it some time before their introduction to CASCON.

Every member of the Pittsburgh group felt considerably better informed as a result of having used these techniques. They felt that the automated services were valuable for their broad coverage even though a high percentage of the relevant alerts came from the manual scanning of the 25 selected journals. Often a computerized service brings to the attention of the user a frequently productive journal which he had not scanned regularly. The broader coverage afforded by the use of computerized services was felt to be more important than any time saved. Most of the group members did not spend enough time in the library to follow up all of their relevant alerts. They checked the most important ones first and the others if they found time. Several of the group members admitted that their selection of relevant hits was influenced by the availability of the reference (others were not held in the library or were in a foreign language). All of the members of the Pittsburgh group said that they were "hooked" on the computer-based services and would maintain them after the experiment if money were available. One could confidently characterize them as being very satisfied with the services.

The reactions of the Harvard group were, on the whole, slightly less favorable. The professors seemed too busy to evaluate the services properly. Two of them delegated practically all handling of their profiles to other members of the group. The group liked the added peripheral coverage but also liked to scan their favorite journals regularly. They were worried about paying for these services, given the depressed state of scientific funding. The research fellows were the most enthusiastic users in the group. One felt that it would be hard to return to manual literature coverage. Another mentioned that the ease of maintaining a bibliography was quite valuable, in addition to the broader coverage in fringe areas. A third, while being very satisfied, noted that a loss of browsing might be a significant drawback. A fourth, a biochemist, felt that the retrieval in well defined areas was excellent, but that more general areas produced more noise. Most of the students, since they were from one research group and working in closely allied fields, felt that one group profile would be worth the cost. The students were generally pleased with the service for the same reasons stated earlier for the Pittsburgh group. The Harvard science librarian was interested in exposure to these services and in seeing them in actual operation.

To summarize, we found that the main advantage of the computerized services was their broad coverage, especially in journals on the borders of a user's primary interest area which he would not usually cover with manual procedures. The services provide an excellent supplementary back-up to manual coverage of key journals in the user's central interest sphere, and provide a valuable means of bibliographic compilation.

XI. USER CHARACTERISTIC PROFILE

In this chapter we have described characteristics which seem to contribute to the successful use of automated current-awareness and retrospective information services by users (see Chapter 7). In Table 19 we indicate these characteristics and our estimates of the probable success of users (on both Current-awareness and Retrospective services) who exhibit them. Probable success is indicated as high (H), medium (M), and low (L). An R entered in a column indicates the level of success for current-awareness service use.

TABLE 19

User Characteristic Profile

	Characteristic	Success H	M	L
1.	Has Ph.D.	CR		
2.	Early professional age	CR		
3.	Eager to use service as soon as available	CR		
4.	Broad information needs	CR		
5.	Longevity in present field of interest	C		R
6.	High professional involvement	CR		
7.	Wants ability to browse the literature			RC
8.	Has access to good library facilities	CR		
9.	Regularly scans journals of interest	C	R	
10.	Research-oriented scientist	CR		
11.	Theoretician		R	C
12.	"Hit-and-run" researcher	R		C
13.	From large industrial concern	CR		
14.	From small industrial concern	R	C	
15.	Applied researcher	CR		
16.	Basic researcher	C	R	

XII. IMPLICATIONS OF BEHAVIORAL RESEARCH

A. Interdisciplinary Nature of the Project

This has been a pioneering endeavor of cooperation between behavioral scientists and chemists within an interdisciplinary effort to establish a chemical information center. The early days of the project were ones of accommodation, where the behavioral scientists had to learn about the general scientific subculture of chemists and their information environment. The chemists on the project's staff had to become acquainted with and sympathetic toward the way behavioral scientists work, i.e., behavioral methodology and research design. Behavioral scientists found that many of their assumptions about the chemist were wrong, and chemists learned that behavioral scientists used different methodologies from those in the physical sciences which are more applicable to the analysis of complex empirical data.

The behavioral scientists found that chemists are often reluctant to spend much time describing in detail how they use the literature of their field. What might be considered an interesting behavioral phenomenon was often viewed as commonplace by the research chemist.

Some members of the behavioral research group entered the project at its early stage with great expectations of doing basic research in cognitive psychology, citation studies, invisible college development, informal communication, and the sociology of knowledge. However, it was determined that these were not directly related to the basic mission of the project and accordingly they were dropped from the group's overall research design. Some of the original motivation for the behavioral study was reduced. In connection with this, an extensive behavioral study designed to use repeated in-depth interviewing of a large stratified user group was a victim of the project's budget crisis. Therefore, a decision was made to settle for descriptive data instead of trying to pursue a research course to develop inferences which were more fundamental to the study of information processing behavior.

The intent of the Behavioral Task Group was to observe the use of new information services, to provide mechanisms for user feedback, and to evaluate the performance of the system with regard to the behavioral implications of the system's design. In the course of meeting these goals, some fundamental research on information processing behavior has necessarily been developed.

In most general terms, the involvement of behavioral scientists on the project was valuable because it encouraged all of the staff, i.e., chemists, system designers, information scientists, and librarians, to be constantly

aware of the behavioral implications of their work. We have tried to high-
light the user's role in the system and point it out unremittingly to others
on the staff. These suggestions have always been well received and acted
upon when possible.

In contrast, concern for the user was almost totally lacking in other
individuals not affiliated with our project. On several occasions at meeting about information systems, we were struck by the lack of meaningful
discussion of the _user_ of information systems and even more by the low
frequency with which the word "user" was even mentioned. In considering
system effectiveness at professional meetings of information scientists,
considerable time was spent discussing hardware, computer times and
costs, abstracting, indexing, formats, keywords, files, inverted files,
marketing, programming, etc. , as though it were preordained that information systems should come into being and that they should obviously be the
most effective way of satisfying the information needs of the scientist.
There was little careful consideration of how or whether the scientist uses
the system.

We have argued for taking individual differences into account in creating
flexible systems which can accommodate the user rather than constructing
inflexible systems to which the user must adapt in order to gain access.
We believe that the user must be kept foremost in the design and operation
of information systems. What might appear to a system's designer as an
orderly and systematic means of handling information may be completely
useless to some scientists. Alternately, what may seem a chaotic and in-
efficient way of processing information to the system designer may work
quite well for some types of scientists.

B. The Need for Behavioral Methodology and Theories

The methodological sophistication of studies on information processing
and system evaluation has been steadily increasing in recent years, but
many problems are left unanswered. More work is needed for deeper conceptualization of test variables and for more efficient research design.
Methodology has, of necessity, been borrowed from the fields of behavioral
science, but information scientists will have to learn how to adapt it to
meet their special problems. The major problem from the point of view of
a researcher involved in this type of study is the lack of support for fundamental psychological and sociological research which is _not_ tied to local
information systems, so that the findings can be generalized to broader
areas of information processing behavior. There is also a lack of theories
which will bring order to the scattered findings of this infant research area.

Advances will come about more quickly as more behavioral scientists
realize the utility of information theoretical strategies for the analyses of

behavioral phenomena. From the early a priori and ad hoc typologies, we expect to see the development of more abstract conceptual schemes with sufficient empirical support to rise above the data from which they were generated.

REFERENCES

1. Chem. Eng. News, July 14, 1969, p. 102.
2. Hays, William L., Statistics for Psychologists, Holt, Rinehart and Winston, New York, 1963: (a) pp. 589-592, (b) pp. 604-606, (c) pp. 493-538.
3. Rosenberg, Victor, Studies in the Man-System Interface in Libraries, Report No. 2, Center for the Information Sciences, Lehigh University, Bethelem, Pennsylvania, 1966.
4. Kline, Ron, Discussion at semi-annual meeting of The Association of Scientific Information Dissemination Centers, Washington, D.C., February 24, 1971.
5. Pugh, Mary Jane, "The Use of Four Current Awareness Systems by an Academic Chemistry Research Group," Master's Thesis, University of Pittsburgh, 1969.

Chapter 4

SYSTEM DESIGN, IMPLEMENTATION, AND EVALUATION

Neale S. Grunstra and K. Jeffrey Johnson

Pittsburgh Chemical Information Center
University of Pittsburgh
Pittsburgh, Pennsylvania

and

Department of Chemistry
University of Pittsburgh
Pittsburgh, Pennsylvania

I. INTRODUCTION

Since 1967, the data processing group of the Pittsburgh Chemical Information Center (PCIC) has had the opportunity to work on a variety of
interesting projects in the information retrieval field [1, 2]. These projects
have included both the implementation and evaluation of several preprogrammed search packages and the design, programming, and implementation of current-awareness, retrospective, and interactive information
retrieval systems. It is felt that these efforts have satisfied the demands
of our users, and that these results can be put to good use on future computerized retrieval projects.

The sections of this chapter roughly coincide with the implementation
of various data bases and software search systems by the PCIC. Since the
project is primarily chemically oriented, the majority of experimentation
has utilized both the CA Condensates and the Chemical Titles (CT) data
bases, both of which are prepared by the Chemical Abstracts Service (CAS).
In June of 1970 CAS announced the availability of a new formating for all of
its bibliographic files. It is the intention of CAS to eventually record all
past as well as future issues of their machine-readable data bases in this
new format. A more detailed description of these data bases and the advantages and disadvantages of each are discussed in the next section of
this chapter.

During the first six to nine months of the project, some project members expressed anxiety to get a system on the air, and yet there was also
a reluctance to plunge into a massive programming effort until the needs
of the users had been clearly defined. To add to the problem, many of the
users were not able to define their requirements more fully until they were
able to review the output of a search system. To resolve this dilemma,
two preprogrammed systems were implemented which would provide the
needed exposure for both the users and the members of the PCIC. Hopefully this approach would not require as large an investment in programming time. First, an effort was made to search the CT data base interactively. This experiment is recounted in the third section of this chapter.
While the experiment was not highly successful, valuable user feedback was
obtained, and increased experience and technology suggest that future interactive search systems may be feasible.

The second effort consisted of the implementation of two retrieval systems operated in the batch-processing mode. One of these systems provided the capability of searching the CT data base. This system is still in
operation. The other system, called the Chemical Abstracts Condensates
Search System (CASCON), provided a retrieval capability of the Condensates file. The latter set of programs was used for nearly one year, until
the beginning of 1970, to obtain feedback information concerning the needs

of users in the scientific community. Approximately six months after the CASCON programs had been implemented, feedback from users of the PCIC's services indicated a need for a number of additional options. Wherever possible, modifications were made to the CASCON programs. However, there seemed to be a number of search options which were impractical to install in these programs without substantial modifications to the system. This indicated a need to either write our own search system, which would include the requested search options, or to implement a more comprehensive prewritten package. The IBM Corporation had indicated at this time that the TEXT-PAC system contained many of the features desired by the PCIC users. Consequently, after soliciting the cooperation of the IBM Corporation in prereleasing TEXT-PAC, a decision was made to implement this system. Plans were to operate the TEXT-PAC system in parallel with the CASCON search system for several months to compare the search options and cost-performance advantages of each system. The TEXT-PAC systems proved to be less expensive to operate for a large volume of users than the CAS system, and they provided more user options. After the initial period of evaluation of the CASCON and TEXT-PAC programs, the CASCON-user profiles were converted to the input format required for the TEXT-PAC programs. This transition took place in the spring of 1970. Current-awareness profiles have been processed using the TEXT-PAC system since that time.

With the impending release of the new Standard Distribution File (SDF) format by CAS, the requirement for a set of new search programs to handle the new file format became apparent. In addition, it was felt that by this time, the programming staff of the PCIC had developed sufficient expertise and exposure to information retrieval systems to undertake such a project. It was decided to develop a system of our own design to search CAS-supplied data bases in the new SDF format. The Chemical Titles, CA Condensates, and SDF search systems are discussed in detail in the Section IV of this chapter.

Enhancement of user service by the implementation of either a retrospective or structure searching capability was also considered. Some of the disadvantages of implementing structure searching are outlined in Chapter 2. There was intense user demand for retrospective searching, so it was decided to implement and test the retrospective search programs which were available as part of the TEXT-PAC search package. The details of this experimental effort are recorded in Section V.

Both because of the poor results achieved in our initial time-sharing effort and also because of the relatively small programming staff available, interactive searching was developed by the PCIC only on a part-time basis. Even so, the programs took on several attractive capabilities. If a programming system of this type had been used in the initial effort to search

CT interactively, the results might have been somewhat more encouraging. Details of the development of this interactive search system are provided in Section VI.

Two computers have been used in the PCIC development effort. Both are part of the main Computer Center of the University of Pittsburgh. The first to be installed was an IBM 360/50 dedicated to time-sharing (hence the implementation of the interactive CT system early in the project). There are currently approximately fifty remote terminals on the campus. The Pittsburgh Time-Sharing System (PTSS) was developed in 1967, and an increasingly reliable time-sharing service has been provided from that time. The second computer, installed in late 1968 and also a 360/50, is dedicated to batch-processing under the full Operating System (OS). The bulk of the computing for the PCIC was performed on the latter system. The configuration has changed with time, and currently consists of 256k bytes of fast-core storage, 1024k bytes of read-only storage, a 2314 disk drive, five magnetic tape drives, a card reader, punch, and high-speed printer. Three intermediate-speed remote job entry stations are also configured to this system, each with a card reader and line printer. The OS software on this computer also progressed in an evolutionary manner from OS/PCP (primary control program) to OS/MVT (multi-programming with a variable number of tasks) with HASP II (Houston Automatic Spooling Program). The batch programs are currently processed using the MVT version of IBM's Operating System. Most of the processing has been done in an over-the-counter environment. Turnaround time varied from two hours to two days.

II. DATA BASE DESCRIPTION

The Pittsburgh Chemical Information Center has conducted both batched and interactive searches on a variety of well-known data bases. However, the project was primarily chemically-oriented, and therefore the majority of experimentation took place using either the Condensates data base or the CT data base, both of which are prepared by CAS. CAS commenced publishing the machine-readable data base, CT[3], in 1962. Subscribers to this service receive semimonthly both magnetic tapes and hard copy issues of titles, authors, and bibliographic citations of current articles appearing in 650 chemical journals. CAS began distributing Condensates in machine-readable format in the fall of 1968. One tape is issued by CAS each week corresponding to the hard copy issues of Chemical Abstracts. Each record on the tape includes an abstract number, title, authors, bibliographic citation, and key words that amplify the content of the article. The abstracts and molecular formulas included in the hard copy of the Chemical Abstracts journal are not included on the tape, although space has been provided in the file to add this information. Each type of record included on the magnetic tape has a specific identifying record number associated with it. Papers from approximately 12, 000 journals are included in this data base.

The Condensates data base is divided into 80 sections. The first 34 sections are published as an odd-numbered issue one week, and the remaining 46 sections are published as an even-numbered issue the following week. The file of 80 sections is divided into five broad divisions: biochemistry, organic, macromolecular, applied chemistry and chemical engineering, and physical and analytical chemistry. The following list provides a more detailed breakdown of the sections of chemistry included in each of these five divisions.

SECTIONS OF CHEMICAL ABSTRACTS

Biochemistry

1. History, education, and documentation
2. General biochemistry
3. Enzymes
4. Hormones
5. Radiation biochemistry
6. Biochemical methods
7. Plant biochemistry
8. Microbial biochemistry
9. Nonmammalian biochemistry
10. Animal nutrituion
11. Mammalian biochemistry
12. Mammalian pathological biochemistry
13. Immunochemistry
14. Toxicology
15. Pharmacodynamics
16. Fermentations
17. Foods
18. Plant-growth regulators
19. Pesticides
20. Fertilizers, soils, and plant nutrition

Organic chemistry

21. General organic chemistry
22. Physical organic chemistry
23. Aliphatic compounds
24. Alicyclic compounds
25. Noncondensed aromatic compounds
26. Condensed aromatic compounds
27. Heterocyclic compounds (one hetero atom)

28. Heterocyclic compounds (more than one hetero atom)
29. Organometallic and organometalloidal compounds
30. Terpenes
31. Alkaloids
32. Steroids
33. Carbohydrates
34. Synthesis of amino acids, peptides, and proteins

Macromolecular chemistry

35. Synthetic high polymers
36. Plastics manufacture and processing
37. Plastics fabrication and uses
38. Elastomers, including natural rubber
39. Textiles
40. Dyes, fluorescent brightening agents, and photosensitizers
41. Leather and related materials
42. Coatings, inks, and related products
43. Cellulose, lignin, paper, and other wood products
44. Industrial carbohydrates
45. Fats and waxes
46. Surface-active agents and detergents

Applied chemistry and chemical engineering

47. Apparatus and plant equipment
48. Unit operations and processes
49. Industrial inorganic chemicals
50. Propellants and explosives
51. Petroleum, petroleum derivatives, and related products
52. Coal and coal derivatives
53. Mineralogical and geological chemistry
54. Extractive metallurgy
55. Ferrous metals and alloys
56. Nonferrous metals and alloys
57. Ceramics
58. Cement and concrete products
59. Air pollution and industrial hygiene
60. Sewage and wastes
61. Water
62. Essential oils and cosmetics
63. Pharmaceuticals
64. Pharmaceutical analysis

Physical and analytical chemistry

65. General physical chemistry
66. Surface chemistry and colloids

67. Catalysis and reaction kinetics
68. Phase equilibria, chemical equilibra, and solutions
69. Thermodynamics, thermochemistry, and thermal properties
70. Crystallization and crystal structure
71. Electric phenomena
72. Magnetic phenomena
73. Spectra and other optical properties
74. Radiation chemistry, photochemistry, and photographic processes
75. Nuclear phenomena
76. Nuclear technology
77. Electrochemistry
78. Inorganic chemicals and reactions
79. Inorganic analytical chemistry
80. Organic analytical chemistry

Condensates has four features that make it more appealing for current-awareness than CT and other CAS data bases. First, since CT includes only 650 journals, Condensates offers significantly greater journal coverage, as well as books and patents. Second, the key word feature is absent in CT. Third, the Condensates user has the option of searching even or odd issues, or both. Thus, if he desires, the user may eliminate wide areas of chemistry, thereby reducing both the number of irrelevant hits and search costs. A final advantage of Condensates over CT is uniformity in hyphenation of words. In CT space has been placed within a word at the point of hyphenation. When a search is being conducted for that particular word, the user is not aware that the word has an additional space. Consequently, this particular term is not retrieved in the search process. This disadvantage has been overcome in the building of the Condensates file.

On the other hand, because the CT data base only includes the title, author, and coden, each record contained on the magnetic tape is much shorter than the corresponding record on the Condensates tape. This means that the CT tape can be searched more rapidly than the Condensates tape. In addition, this data base is available from January, 1962 to the present, making it a more desirable retrospective file.

In July, 1970, CAS announced a new tape format entitled the Standard Distribution File (SDF) format. Presently CAS is supplying Condensates in both the old and new formats. At some future date only the new format will be available. This new format will eventually apply to all of the data bases which are currently being produced by CAS. This would have the advantage of standardizing the data which are on the files produced by this organization (Condensates, Post-P, Post-J, CBAC, CT, and BJA). The advantage of this approach is that one search system can be used to search all of these data bases, significantly reducing program development costs for the user. Briefly, each data element (e.g., author, title, date, etc.)

is identified in the new SDF [4, 5] data base by the following: a unique data element number which identifies the data element, a storage mode indicator (binary, decimal, hexadecimal, octal), a number which indicates the length of the data element, and the value of data element. The new data base format provides more flexibility, in that all types of data elements are assigned a unique identifying number. In former data bases several data elements were combined, making selective searches difficult. Also, new data elements can be added to the file with few programming changes, merely by adding a new identifying number to the file. The new format also permits data to be recorded in upper and lower case.

The data base has been recorded in the American National Standards Institute (ANSI) character set to conform with recent government standardization requirements. In order to process this data base on an IBM 360 computer system, the data base must be converted into the Extended Binary Coded Decimal Interchange Code (EBCDIC) character set. Similar procedures will be required when processing the SDF data base on computer equipment which are EBCDIC-oriented. The SDF format is a serial file. This means that the potential CA Condensates user must either search the file in a linear fashion, which is time-consuming and costly, or else convert the data base into a more efficient search format, which is also expensive.

Table 1 contains sample records from the CT, CA Condensates, and SDF files.

TABLE 1

Printout of Chemical Titles, CA Condensates, and SDF Format Files

_____

Chemical Titles

ABMGAJ-0025-0213 11 ZAHN D   KLINGER R   BROX D
ABMGAJ-0025-0213 12 HERMANN M WACHTER R WACHTER E
ABMGAJ-0025-0213 21 HEXOSE MONO PHOSPHATES IN THE NORMAL
                          LIVER. =
ABMGAJ-0025-0213 31   70   213-23   CT7102

ABMGAJ-0025-0225 11 BLUTH R   BANASCHAK H
ABMGAJ-0025-0225 21 INFLUENCE OF ALPHA AND BETA RECEPTOR
ABMGAJ-0025-0225 22   BLOCKING AGENTS ON THE OXYGEN CONSUMP-
ABMGAJ-0025-0225 23   TION OF MET HEMOGLOBIN CONTAINING
                          ERYTHROCYTES. =
ABMGAJ-0025-0225 31   70   225-32   CT7102

CA Condensates

084232H	7019	5	M PROTEINS MEMBRANES REVIEW*
084232H	7019	5	AMP DNA MEMBRANES REVIEW*
084232H	7019	5	DNA AMP MEMBRANES REVIEW*
084232H	7019	5	MEMBRANES DNA AMP REVIEW*
084232H	7019	5	REVIEW MEMBRANES DNA AMP*
084233J	7019	1	PRPCBSOO 1, 000068 03470405**
084233J	7019	4	PROGR. PHYTOCHEM. *
084233J	7019	2	BIOCHEMISTRY AND PHYSIOLOGY OF PHYTOCHROME. ==*
084233J	7019	3	FURUYA M. *
084233J	7019	4	(UNIV. NAGOYA, NAGOYA, JAPAN)*
084233J	7019	5	PIGMENTS PHYTOCHROME PLANTS REVIEW*
084233J	7019	5	PHYTOCHROME PIGMENTS PLANTS REVIEW*
084233J	7019	5	PLANTS PIGMENTS PHYTOCHROME REVIEW*
084233J	7019	5	REVIEW PIGMENTS PHYTOCHROME PLANTS*

SDF Format

00127001	00008	00000074
00547001	00009	69280061A
00557001	00006	CPROAI
00597001	00022	Thorne, J. Geoffrey M.
005B7001	00043	Postal physicochemical measurements service
005D7001	00023	Chem. Process. (London)
005E7001	00006	000069
005F7201	00002	15
00607001	00010	10) (Suppl.
00617001	00008	S4-S5, S7
00637201	00002	EN
00637302	00003	Eng
00667001	00014	CA07217086147Z

III. THE PTSS—CT EXPERIMENT

Early in 1967 the Library Committee of the Chemistry Department at the University of Pittsburgh decided to initiate a computer-based, inter- active, current-awareness service. The data base chosen was CT. CAS provided in IBM 360-compatible Basic Assembly Language (BAL) program to search this data base.

The computing facilities at the University of Pittsburgh at this time included an IBM 360/50 dedicated to time-sharing. The Pitt Time-Sharing System (PTSS) [6, 7] was still in the shake-down period. The CAS program was designed for OS which had not been implemented yet at the University. Attracted by the interactive features of a PTSS-CT current-awareness system, the Library Committee decided to have the search program for PTSS rewritten. This was accomplished by Mr. Griffith Smith, a chemistry graduate student. All of the logic and truncation features of the CAS program were retained. In addition, nested parentheses were allowed, so that several questions could be coded in one user profile. It was designed to be used by the research chemist in an interactive mode. Therefore, only one profile was searched per pass of the tape. The user could develop and refine his search strategy and profile at the remote terminal.

The results of this early interactive current-awareness experiment have been described by Bloemeke and Treu [8]. In short, the experiment failed. One major problem was the instability of this early version of PTSS. The system was designed for short student jobs, and CT searching degraded the system significantly. Also, search times were too lengthy to be "interactive." The user had to be sophisticated as well as patient. He had to load the appropriate CT tape and the search program, correct his profile if necessary, and monitor the search process.

Nevertheless, searches were conducted this way until late 1968, when the University rented a second 360/50 and dedicated it to batch-processing under OS. The CT current-awareness service has continued in the batch-processing mode.

The reader should keep in mind that this effort was an early attempt at interactive searching. Today, due to the increased reliability of PTSS and to increased programming expertise, an interactive CT system might be feasible.

Some factors which may be considered are the searching of the file from disk rather than from tape; inversion of the file; removal of large numbers of blank characters from within the record; and elimination of searching duplicate data elements and unnecessary data elements from within the records, e.g., the citation number, page numbers, and volume number. An enhanced search package could also selectively search either the author or the title information. However, there are still some questions to be resolved relative to the feasibility of searching a current-awareness data base interactively. Is there adequate high-speed random-access storage available? Is the time-sharing system reliable enough to support a relatively long information-retrieval job without degradation of service to other users? Will the user be willing to learn the command language and direct the search from his terminal? Each potential user group will have to make the determination as to the practicality of an interactive current-awareness service.

IV. CURRENT AWARENESS

A. Chemical Titles (CT)

When the IBM 360/50 with OS became available, batch CT search programs were implemented. This took place late in 1968. Because of the shortness of each CT tape record, a serial search technique was employed. It was felt that the computer time spent in reformating the CT file would offset any savings in improved search efficiency. Due to this constraint, the CT search has remained essentially the same as it was in 1968-69. Little was done to streamline the programs. The system is still in use, with about 90 active users (as of March, 1971). The only proposed modification is to place the output alerts on cards. Indicative processing costs for the CT file have been included in Table 2. These were derived using an IBM 360/50 with 128k of core, running under OS/PCP.

The PCP version of OS does not provide statistics for CPU processing time. The execution times shown in Table 2 therefore include such variables as set-up time, CPU time, and the printing of the alerts. This may account for the irregularity in the costs per profile. Nevertheless, it can be safely stated that the cost per profile of processing one issue of CT would be less than $2.00 per profile for a profile containing approximately 18 terms. Table 2 reflects timings for production searches of representative issues of the CT data base. Variance in the cost per profile could also be due to the slight differences in the number of citations recorded on individual issues of CT.

B. Condensates

Shortly after CAS announced the availability of Condensates, users indicated a preference for this new data base. This preference was due primarily to the differences in the data base as noted in Section II of this chapter. In order to provide a Condensates search capability as soon as possible, it was decided to utilize a set of programs supplied by CAS to search the Condensates file. This approach has the advantage of allowing the PCIC to obtain valuable feedback from users without incurring the tremendous costs required in writing our own system.

By the winter of 1968 the CASCON current-awareness services were available on a production-processing basis. This service has built up to a current level of approximately 200 users of the CASCON system. Several limitations of this system motivated the PCIC to investigate the possibility of either obtaining another search system or designing one of our own.

TABLE 2

Chemical Titles Processing Costs

Number of profiles	Number of terms	Number of terms/profile	Alerts	Number of alert/profile	Execution time	Total cost ($)	Cost/profile
48	863	18.0	882	18.4	25.3	84.97	1.77
59	1070	18.1	1337	22.7	34.5	116.06	1.97
79	1423	18.0	1804	22.8	41.1	102.74	1.30
87	1485	17.1	2513	28.9	51.1	170.39	1.96

Several prewritten systems were reviewed and rejected, because of excessive costs, complexity, or limitation of capabilities. The PCIC was able to obtain a prerelease copy of the TEXT-PAC programs [9] from IBM. These programs were written by an IBM user and made available for distribution to IBM customers. IBM modified the TEXT-PAC system to search technical information on an in-house basis, and particularly to search the Condensates data base.

While some program debugging would be required on the part of the PCIC programming staff, it seemed that this might amount to a significantly smaller effort than would be required to write an entire system. In addition, the TEXT-PAC system seemed to contain most of the search options requested by the users. These included expanded search logic capabilities which were not available in the CASCON program (see Table 3). The TEXT-PAC system also had a variety of output formats, including 80-column cards and 3 x 5 cards, as well as the regular computer stock paper. It provided the user with additional statistical information, including a KWOC (Key Word Out of Context) index, word frequency distribution, and the capability of flagging terms which caused an alert. A file-reformating program, which inverted the file alphabetically within word length, held some promise for reduced processing costs. TEXT-PAC had a program which would keep an accounting of the alerts each user received. This provided an ideal opportunity to start a computerized feedback capability. This program would record the alerts received by each user onto a computer-readable magnetic tape. As the user indicated to us which alerts were relevant and which were not, this information could be stored on a separate tape file and a program written to calculate statistical information as to the percentage of relevancy and nonrelevancy. Finally, a complete set of retrospective search programs was available as part of the TEXT-PAC system.

On the other hand, the TEXT-PAC system had some drawbacks which made the decision a difficult one. For example, the TEXT-PAC system, because of the file reorganization technique, precluded the use of left-hand truncation. It was felt that this might be a serious disadvantage, particularly in handling complex organic chemical nomenclature. Further, the TEXT-PAC system does not have the capability of weighting search terms in the profile.

In spite of these potential drawbacks, the advantages seemed to outweigh the disadvantages, and it was decided to implement the TEXT-PAC system and process various profiles on an experimental basis while the CASCON system was used to process the user profiles on a production basis. In this manner, the advantages and disadvantages of a TEXT-PAC system could be evaluated. This also gave us the opportunity to evaluate the TEXT-PAC and the CASCON system from a cost/performance point of view. This was done during the latter half of 1969.

TABLE 3

Summary of the Capabilities of the TEXT-PAC and CASCON Systems

Description of features	Available with CASCON system	Available with IBM TEXT-PAC systems	SDF search system
SDF sectioning program available[a]			X
Profile update program available		X	
User statistical information			
a. Word frequency distribution		X	X
b. Alerts per profile	X	X	X
c. Total number of alerts	X	X	X
d. Total records scanned		X	X
e. Flag terms which caused alert		X	
f. KWOC index		X	
Edit program capabilities[b]			
a. Check profile logic		X	X
b. Check profile content	X	X	X
c. Print profiles		X	X
Output capabilities			
a. Bibliographic citation	X	X	X
b. 80 column cards, 2-up		X	X
c. 3 x 5 cards		X	
Program available for feedback for user information		X	
Profile composition capabilities			
a. Left truncation	X		X
b. Right truncation	X	X	X
c. AND/OR logic	X	X	X
d. Absolute		X	
e. NOT logic	X	X	X
f. WITH logic		X	
g. Option to search by			
Author	X	X	X
Title and key words	X	X	X
Dates		X	
Corporate author[c]		X	X
Locations		X	X
Journal coden	X	X	X

TABLE 3 (Continued)

Summary of the Capabilities of the TEXT-PAC and CASCON Systems

Description of features	Available with CASCON System	Available with IBM TEXT- PAC systems	SDF search system
h. Weighting capability	X		
i. Nested logic capacility		X	X
j. Adjacent logic	X	X	X
k. Security check feature (for confidential information)		X	
l. Upper and lower case capability		X	X

[a] Available in retrospective search.

[b] The CASCON edit program is part of the search program. This means that all acceptable profiles must be searched. The TEXT-PAC system by contrast, allows the edit programs to be run independently of the search.

[c] SDF search system allows user to specify the number of corporate authors to be searched.

At the time this experiment was conducted, the IBM 360/50 computer dedicated to batch-processing was running under the PCP operating system. Shortly after the completion of the experimental phase using TEXT-PAC, the computer center updated the operating system to MVT. Had the experiments been run using the MFT operating system, the timings would have been quite different, although the overall results would not have changed. The details of the experiment which compare the TEXT-PAC and CASCON search systems were reported in the November, 1970 issue of the Journal of Chemical Documentation*.

CAS supplies the CASCON search system without charge to CA Condensates tape subscribers upon request. The Chemical Abstract Service considers the CASCON system to be only a basic set of programs to be used with the CA Condensates tape. The source programs are in Basic Assembly Language (BAL) and contain approximately 7,000 cards. The search

---

*The following material is an excerpt from Ref. 10.

program (plus Operating System) requires approximately 77k bytes for storage of the object program, and will utilize all available core when loading search terms into memory.

A flow chart of the CASCON system is given in Fig. 1. Input to the programs include an issue of CA Condensates on tape and the interest profiles of the users. The program edits the profiles for validity, creates a profile table in the computer memory, and searches the entire tape, character-by-character, matching characters on the CA Condensates tape with entries in the profile table. Options must be specified in the search profile which permit the user to search author, coden, and/or title and key words. Records on the tape that match a given profile term are called alerts. The alerts for each profile are then sorted into sequence by question number, weight (i.e., significance of each alert as determined by the number of matches encountered between the document and profile), and digest number. The sorted profiles are then printed and distributed to the user.

IBM has versions of TEXT-PAC which are compatible with either IBM 7090 or 360 series computers. TEXT-PAC represents a more general information retrieval system than CASCON. A conversion program is needed to translate the data base to be searched into the required TEXT-PAC input format. The source decks contain approximately 16,000 cards and the programs require 128k of memory (the conversion program is the only program which requires 256k).

Flow charts of the TEXT-PAC system are shown in Fig. 2, 3, and 4. For clarity, many of the TEXT-PAC options have been eliminated from the flow charts. However, these options have been itemized in the summary of the capabilities of both systems (Table 3).

The conversion of the CA Condensates data base into TEXT-PAC search format is illustrated in Fig. 2. Two tapes are produced as output of the conversion subsystem: 1) a searchable TEXT-PAC data base, where each word in the data base is sorted alphabetically by word length; and 2) a condensed text tape, in readable context, used by the program which reformats the TEXT-PAC output (TRCO13).

Figure 3 describes the handling of profiles. The profile update program (TRC001) allows the user to store profiles on tape or disk and to add, change, or delete selected items as desired. The profiles are edited, listed, and sorted alphabetically within word length.

As shown in Fig. 4, these profiles and the converted CA Condensates data base are the input to the search program (TRC011). The resulting alerts are sorted according to the profile number, merged with the condensed text data, printed, and sent to the user.

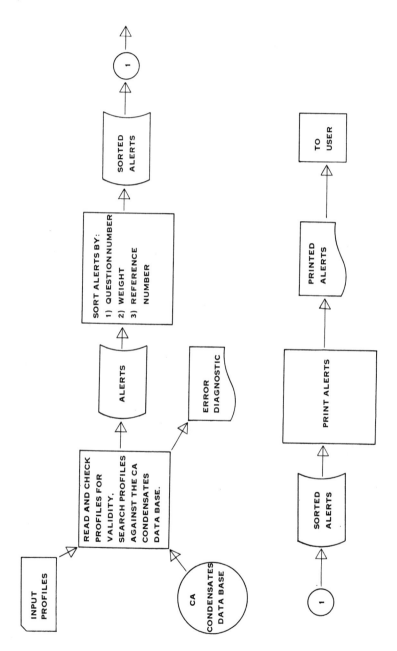

FIG. 1.   Flow chart of CASCON processing.

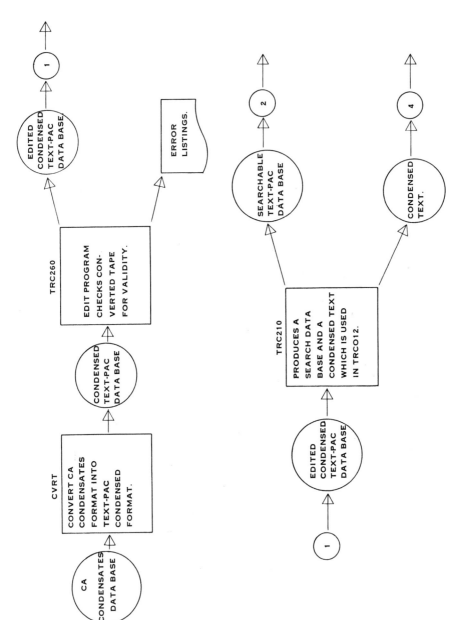

FIG. 2. Flow chart of <u>CA Condensates</u> to TEXT-PAC data base conversion.

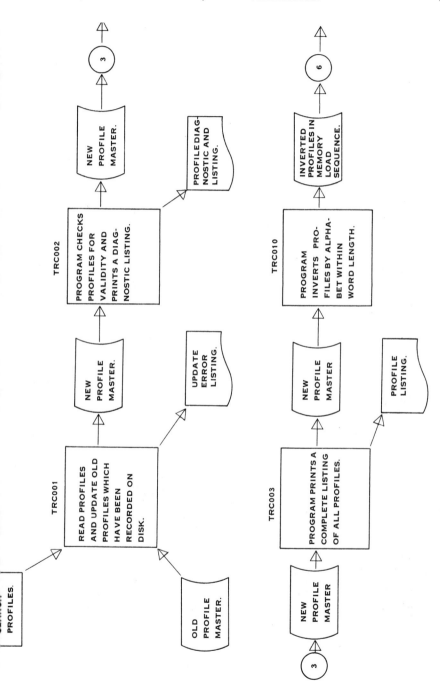

FIG. 3. Flow chart of TEXT-PAC profile preparation.

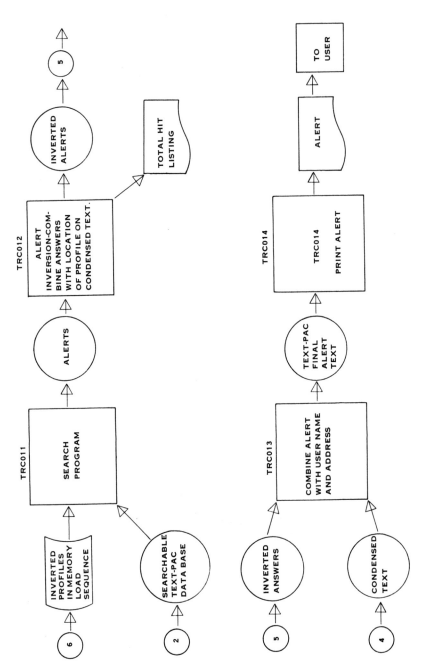

FIG. 4.   Flow chart of TEXT-PAC search processing.

Table 3 summarizes the capabilities of both the CASCON and TEXT-PAC search systems. In general, the TEXT-PAC system seems to provide the user with more options, from the standpoint of profile logic as well as program capabilities. TEXT-PAC allows the user to specify nested logic. TEXT-PAC can also selectively search a wider variety of records on the data base, including searches by date, location, and security classification. TEXT-PAC also provides variety in the formating of output, allowing for 3 x 5 cards, bibliographic citations, and 80-column continuous form cards, printed two cards across.

Auxiliary reports are also available in the TEXT-PAC system, such as a word frequency report which shows the total number of occurrences of each word in the data base as well as a count of the number of documents in which each word occurs. A dictionary compare program allows the TEXT-PAC user to check the correctness of spelling of words in the data base.

[Cost/performance experiments were conducted to determine whether it would be more advantageous to use CASCON or TEXT-PAC. The details of these studies follow.]

Both IBM and CAS provided the search programs on a program tape in source and object format, accompanied by documentation providing instructions on how to implement the system. The documentation supplied with the TEXT-PAC system (although in draft form) was more complete than the documentation supplied with the CASCON system. TEXT-PAC was received as a prerelease of a type-three program from IBM. The availability of the TEXT-PAC system to a prospective user under the IBM "unbundling" policy is still to be resolved. Both the CASCON and the TEXT-PAC systems required two to three months to implement. This time, however, could have been reduced considerably had it not been necessary to work in an over-the-counter environment in the University of Pittsburgh Computer Center. Data set names used by the organizations who wrote the original programs had to be changed to correspond to those familiar to the Operating System of the University of Pittsburgh. This presented a potential hazard in that, if a data set name is mispunched during the transition, no OS diagnostic is given, yet the program will not run correctly, and considerable time can be devoted to locating this problem. The CASCON print program was modified to record the citations that are normally printed on to a magnetic tape (7 track, 556 BPI). This tape is then printed off-line on a small IBM 1401, available at a substantially reduced charge, rather than printing the tape on the 360/50. This has reduced processing costs substantially (15-25%). A similar modification was completed for the TEXT-PAC system.

Further, this program was written to print the output on continuous form cards, two cards across, rather than on stock paper. Identical information is printed on two cards, both of which were distributed to the

user. The user is requested to return one of these cards after completing
some questions, preprinted on the card, concerning relevancy. This information will be read back into the computer for feedback studies. This
type of output proved useful to the members of the PCIC involved in user
feedback studies and also provided the user with a more easily handled (and
filed) citation. Also, additional coding has been added to the CASCON programs to provide the user with key words as part of the printed output.
This capability was available with TEXT-PAC upon receipt of the system.

Both TEXT-PAC and CASCON have been modified to cause all processing messages to print on the printer rather than on the computer console. This permits all processing messages to be printed on one device
in a chronological manner, which is helpful in debugging. This is particularly important in an over-the-counter processing environment where the
user does not have hands-on access to the computer or to the console log.
Service to users has been improved by separating the user profiles according to odd-even tape preference. Prior to this all profiles were combined
and processed against both the odd and even issues of Condensates.

Some explanation is required to describe the variables relating to the
analysis of the two systems. The input profile format of each search system differs. As a consequence, a program was written to automatically
translate the Condensates profiles into TEXT-PAC profile format. This
assures as nearly as possible that the same profiles (and consequently
similar search logic) will be used by both search systems. Since TEXT-
PAC does not handle left-hand truncation, this option was removed from
both the TEXT-PAC and CASCON input profiles.

Wherever possible, programs which are not available in both systems
have been eliminated in making this comparison. For example, one of the
TEXT-PAC edit programs prints a listing of the profiles. This program
is not available in the CASCON system and therefore was eliminated from
the timings altogether. In addition, the print timings for both systems
have been eliminated for three reasons: to reduce test costs; because the
print costs would be fairly constant for both systems; and most of the
printing for production runs would be handled by an off line printer.

To establish a constant data base for both systems with the same number of input records, the same issue of CA Condensates (Volume 70, No.
16) was used as input for all tests. The CA Condensates issue contains
5,413 records. The reformatted TEXT-PAC issue contains 4,992 records.
The difference in the number of records is due to the fact that TEXT-PAC
converts the CA Condensates file into a more efficient TEXT-PAC search
format. During this process, TEXT-PAC eliminates any records which
are not in acceptable format. This procedure is currently eliminating
approximately 10% of the original file which, of course, is a serious factor
as far as relevancy is concerned. This difference in the number of abstracts

available to the search programs will also affect the processing time in favor of TEXT-PAC. This factor has been normalized in all timings included in this chapter. Improvements have been made subsequent to the timings presented in this study which will eliminate this problem.

The CASCON search program uses a character-by-character search. Generally, this type of search has proven to be extremely slow in processing. This is particularly true when one considers that the data base being searched is serial in nature. The profile edit (checking) program in the CASCON system is built into the search program. This means that it is not possible to edit the profiles without going directly into the search phase. With such an arrangement, profiles which are in error could not be corrected prior to the search phase. Under the CASCON system, erroneous profiles are merely ignored, or if serious discrepancies arise, the search is cancelled. This proves to be a costly arrangement from the point of view of machine time.

The TEXT-PAC system has done away with many of the disadvantages outlined above. For example, TEXT-PAC searches are run on full words rather than on a character-by-character basis. This is done through the CA Condensates—TEXT-PAC conversion programs which put the CA Condensates data base into TEXT-PAC format and at the same time inverts the terms in each document in the data base alphabetically within word length. This procedure speeds up the search significantly as the search program merely looks through the list of words in each document of the same length as the term specified in the profile, rather than searching all terms in the document. At the same time, TEXT-PAC does search all of the data in the document rather than searching only selective fields. However, the structure of the inverted file precludes the availability of left-hand truncation in the search profile (right-hand truncation is available). Another disadvantage of the TEXT-PAC system is the absence of weighting of the search terms.

Table 4 shows the processing costs for various quantities of input profiles for each system. These results have been obtained using an average of twelve search terms per input profile. The term "cost per profile" is calculated by dividing the total processing costs by the number of alerts received as output from each set of programs.

It is interesting to note that, while costs per profile decrease for larger input volumes, costs per alert have increased slightly for both the CASCON and TEXT-PAC systems. It is felt that this slight increase in cost per alert is due to the reduced number of alerts produced compared to the increase in the number of input profiles. For example, the average number of alerts retrieved (TEXT-PAC) when processing 120 profiles was 23.4 per profile, while the number of alerts retrieved at the 180 profile level was only 17.2, and 12.3 for 270 profiles. The average number of alerts retrieved per profile decreases in the same manner for the CASCON system.

TABLE 4

TEXT-PAC and CASCON Comparative Processing Costs

Number of profiles	TEXT-PAC						CASCON		
	Excluding tape conversion costs			Including tape conversion costs					
	Cost per profile($)	Number of alerts retrieved	Cost per alert($)	Cost per profile($)	Number of alerts retrieved	Cost per alert($)	Cost per profile($)	Number of alerts retrieved	Cost per alert($)
30	1.703	85	.601	5.336	85	1.883	1.623	866	.056
120	.745	2868	.031	1.653	2868	.069	1.110	N/A	N/A
180	.625	3094	.036	1.231	3094	.072	1.004	3963	.046
270	.556	3342	.045	.960	3342	.078	1.029	4899	.057

It should be noted that these timings are for an even issue of <u>CA Condensates</u>. It is expected ,that the cost per profile would decrease similarly for an odd issue, but the cost per alert may change due to the difference in the subject matter.

Timings were conducted for 30, 120, 180, and 270 input profiles for both the CASCON and TEXT-PAC systems. These results are presented in Figure 5. While multiple timings were conducted for each input level, the lowest values have been presented here, using the rationale that, since the processing was performed in an over-the-counter environment, the only variation between timings at a given volume of input profiles would be the result of set-up time (since the content of the profiles and data base remained constant).

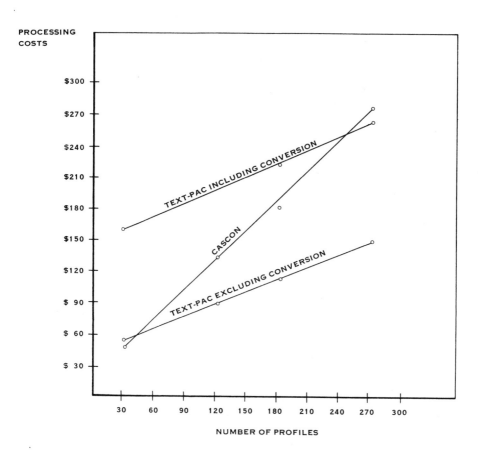

FIG. 5. Comparison of CASCON/TEXT-PAC processing costs

Two sets of costs have been used for TEXT-PAC; one including the processing costs required to convert the CA Condensates tape into TEXT-PAC search format, and the other excluding conversion costs. Both sets of figures have been included because the conversion processing generates TEXT-PAC searchable tapes which are also usable in the TEXT-PAC retrospective search system. Hence, the costs for conversion processing could be added to the current-awareness costs if no retrospective search processing is anticipated by a prospective user of the system, included with retrospective costs, included on a pro rata basis in the processing costs of both the current-awareness and retrospective systems.

Obviously, cost justification of TEXT-PAC compared to CASCON depends upon the distribution of TEXT-PAC conversion costs. This can be seen from Fig. 5. Excluding the conversion costs, the break-even point is in the area of 240 profiles. Splitting conversion costs between both TEXT-PAC current-awareness and retrospective systems places the break-even point at approximately 140 profiles. Costs per input profile as well as costs per unit of output (alerts) are indicated in Table 2. TEXT-PAC looks more attractive cost-wise either where a user anticipates continually large volumes of input, or where conversion costs can be amortized through retrospective processing.

Costs, however, do not tell the complete story. TEXT-PAC seems to have a wider variety of user options available in the system which, to some extent, may serve to offset small differences in processing costs. In the TEXT-PAC system, for example, an optional word frequency report is available which indicates both the total number of occurrences of a word in the data base as well as the total number of documents in which each word occurs. This type of information can prove useful in the preparation of input profiles. TEXT-PAC also provides greater variety in the formating of printed output.

C. SDF Search System

In January, 1970 CAS announced their intention to release by June, 1970 a new file format for the Condensates data base. This file would be distributed in parallel with the old format until some time in 1971. The date was to be determined after all Condensates users had sufficient time to develop new search capabilities. The PCIC took two approaches in processing this new file.

First, a program was written which permits the reformating of the SDF into the file format required for the TEXT-PAC system. Once the SDF is reformated, the resulting output tape is then usable as input to the TEXT-PAC search system. In addition, since most other CAS data bases are being made available in the SDF format, these data bases could also be reformated into the TEXT-PAC search format using this conversion program. The conversion program, therefore, has expanded the capabilities of the PCIC to search all of the files made available by CAS in the SDF format. Also, the outputs of the search system are consistent, because all of these files will be processed using the same information retrieval system — the TEXT-PAC system.

We believe the new SDF search system is more efficient than serial character-by-character searching, and yet does not require the costs and complexities of complete file inversion. The rationale behind the SDF search system is as follows. In order to search the SDF tape on an IBM 360 computer, the file, which is in ASCII format, must be translated by the computer into EBCDIC format for processing. While this is taking place, each word in the data base is examined and the length for each word is inserted in the data base preceding the word. This numeric value is called a length modifier. Length modifiers have also been added to each of the search terms in the profiles which are loaded into the memory of the computer. A preliminary search compares the length modifier and first character of a word in the data base to the length modifier and first character of a word from a user profile. The computer is able to skip over any terms in the data base which are not of the same length and first character as the term in the search query. This eliminates the searching of unnecessary terms in the data base. In addition, this type of file modification is not as complex or time-consuming as a full inversion of the file. The addition of length modifiers to the SDF file takes place in the same time required to read the file to translate from ASCII to EBCDIC - hence, there is effectively no time required for file conversion.

For example, if a data base consisted only of the words, "THE BILL OF RIGHTS," the program translating this information from ASCII to EBCDIC would add length modifiers in front of each word with values of 3, 4, 2, and 6, respectively, as follows: "3THE4BILL2OF6RIGHTS." If a query were entered requesting all material containing the word "RIGHTS," the search program would add the appropriate length modifier to the word "RIGHTS" (in this case, "6RIGHTS"). The pointer in the search program would initially be set to the value "1," to point to the length indicator for the first word in the data base (in this case, "3"). Since the length modifier and first character for the query term (6R) is not equal to the length modifier and first character of the term in the data base (3T), the term in the data base is not one which was requested by the user. The value of the length modifier associated with the term in the data base (3) is added to the original value of the pointer (1), setting the pointer to a value of 4, which is the relative position of the next length modifier in the data base. In the same manner, 6R would then be compared to the next length modifier and first character of the next term in the data base, or 4B. Using

this technique, all of the words in the data base would be skipped until a
word of equal length were encountered, in this case the word "RIGHTS, "
which would represent a satisfactory answer to the query. Hence, no
character-by-character searching is conducted until a correct-length
modifier is encountered and until the first characters of query term and
data base term are equal. At the time of this writing, it was not possible
to conduct a full series of comparative timings. However, initial searches
using the SDF format tapes indicate that the search times are at least
competitive with the TEXT-PAC system. The new system is programmed
for a batch-processing environment, similar to the other information
retrieval systems discussed thus far. It is written in BAL, requires only
65k of core, runs on an IBM 360/30 or above, and requires tape and disk
storage. Since this system processes the SDF tapes, it will also be able
to search all of the other tapes made available by the CAS in SDF format.
Logic capabilities available in the new SDF search system compared to
the logic capabilities of both the CASCON and TEXT-PAC system are
presented in Table 3. A more complete documentation package is
available [11, 12].

V. RETROSPECTIVE SEARCHING

Users of the PCIC indicated a strong desire for retrospective
searching services. The TEXT-PAC source program tape contains a
series of programs which were designed to provide this capability. These
programs have many characteristics in common with the TEXT-PAC
current-awareness information retrieval systems. Perhaps the most
significant difference between the current-awareness and retrospective
search systems is in the file organization. The current-awareness
system converts the Condensates tape into an inverted file organized
alphabetically within word length. When a citation matches a search query,
the abstract number of the inverted tape is retrieved. This abstract
number is then passed against the original data base to select the full
citation. This is done because the inverted data base is not in a readable
form. In the retrospective search conversion programs, the textual
information (representing data which is in the correct context) is placed on
the same file following the inverted data. This eliminates the need to
have the TEXT-PAC retrospective system read a separate text tape to
select the full citation, since the in-context information is already on the
retrospective file. This feature permits the retrospective file to contain
duplicate citation numbers, and in addition, does away with the need for
mounting multiple textual tapes to pull off the in-context citations.

The retrospective search system requires approximately 256k of core,
is written in BAL, and is programmed on approximately 10, 000 source-
language punched cards. This system took approximately three man-

months to implement in an over-the-counter processing environment. In addition to printed citations, two of the key reporting features available in the TEXT-PAC retrospective search system include a statistical report, which indicates those terms in the citation which caused the citation to be selected as an alert. An additional output option of the retrospective system allows the user to have the citation numbers punched on cards. After reviewing the terms which caused a citation to be selected as an alert, the user has the option of either resubmitting the card containing the citation number for printing of the citation, or discarding the card if the citation is irrelevant.

In addition to implementing the retrospective search programs, it was also necessary to build a retrospective data base from the weekly Condensates tapes. These tapes have 34 subdivisions or sections recorded on the odd-numbered tapes, and 46 sections recorded on the even-numbered tapes (see pp. 89-91).

Initially it was believed feasible to build a retrospective data base containing one section, for example, only Section 1 (History, Education, and Documentation) of the CA Condensates file. This approach then would require extracting and merging Section 1 from each current-awareness Condensates tape onto a retrospective data base. This would result in 80 retrospective files – one for each section. This approach was eliminated because too much relevant information in other associated sections of the file would be excluded from the data base. Also, an excessive amount of processing would be required to make separate data bases for each of the 80 sections. An alternative method which significantly reduced both of these disadvantages was developed. This method would extract all the sections under Division 1 (Sections 1 through 20) from the current-awareness data base onto a retrospective data base, thereby including all 20 biochemistry sections. This method was chosen for the following reasons: sufficient material was included in the retrospective data base to provide satisfactory recall for the user; unrelated material from other divisions of the Condensates file was removed from the retrospective data base; significantly less processing time would be required to build this retrospective data base; and a survey of the chemists using the current-awareness retrieval system indicated that a data base built in this fashion would be of significant interest to them.

Preliminary tests of an early version of this data base seemed to support the advantages outlined above and also indicated that processing costs would be inexpensive. Initially, it was decided to build a retrospective-search data base using only the organic chemistry division for the most recent 12 odd-numbered tapes, representing six months' data. The searches could then be run against this test data base to prove the feasibility of the approach described above. If this seemed practical, a decision would then be made to go ahead and build data bases for the other four divisions of the Condensates files. On the other hand, if this approach was impractical, it would not have been necessary to spend the additional funds to build data bases for the other four divisions. The total costs using an IBM 360/50 with OS/MVT and HASP II at $150 per hour were as follows. Merging each Division of a Condensates current-awareness data base into the retrospective file averaged about $15 per tape. For convenience, retrospective files were constructed to contain only six months' data. Each retrospective file had to be reformated into the TEXT-PAC search format. This cost an additional $300. Searches using this sectioned data base have proven to be fast and economical, providing the user with fewer irrelevant citations.

Costs for processing the initial retrospective searches are included in Table 5. The first four entries represent some preliminary test runs which were made using the TEXT-PAC retrospective search system and a Condensates data base constructed as outlined above. The data base consisted of six months' worth of merged Condensates tapes, organic division. The CPU time required to process from one to 20 profiles ranges between 1.97 minutes and 22.93 minutes. Processing costs for a single profile search amounted to $28.54. The total cost for the processing of 20 profiles, including the punching of cards for a statistical report, came to $90.76. The search costs per profile dropped steadily as the total number of profiles submitted to the search system increased. The cost per profile dropped to $4.03 per profile when 20 profiles were submitted to the retrospective search system. It was estimated that a realistic batch size for retrospective searching by PCIC users would be between the 12 and 20 profile level. The per-profile cost for searching six months of data averages about $4.50. If the data base had not been sectioned, costs would have been substantially higher. Based on these preliminary cost figures, the other four divisions of the Condensates files were also merged into retrospective data bases.

A comparative run was conducted using the CT programs and six months' worth of merged CT input. As indicated in Table 5, 20 profiles were submitted against this data base, and total cost amounted to $90.04, or $4.50 per profile. The processing costs for searching CT retrospectively using the original CT programs compares favorably with the TEXT-PAC retrospective search system. This is due to the fact that the CT records are substantially shorter than the records on the Condensates data base, requiring less search time. It further indicates that, because of the shortness of each CT record, there is little benefit to be derived in reformating the data base into a more elaborate format such as that required by the TEXT-PAC system.

Some additional experiments were conducted using the newly constructed retrospective data bases to determine the effects on the overall search
times of varying the number of profile terms, output citations, and size of
the data base. The results of these timings are presented in Tables 6 and 7.
These timings were compiled using the TEXT-PAC retrospective search
system in conjunction with a retrospective data base constructed for one of
the five divisions of chemistry, as outlined previously. The programs were
run on an IBM 360/50 with 256k bytes of storage, under OS with MVT and
HASP II. Table 6 illustrates the CPU time as a function of the number of
terms in the profile. In short, all other variables (size of the data base
and number of output alerts) were held constant. CPU processing time
would then be strictly a function of the number of input terms. Multiple
timings were conducted for each entry on the table.

The number of alerts which are retrieved also affect the processing
time. In Table 7 the average number of input terms has been held constant
while varying the number of alerts received from the search. The results
presented indicate that the retrospective-search processing times are dependent upon the number of alerts to be recorded (and therefore also upon
the speed of the I/O devices).

It can be seen that the TEXT-PAC retrospective search processing
time is a function of at least three variables: the number of terms per
profile; the number of alerts; and, of course, the number of citations on
the data base being searched. Consequently, search times can be improved
by optimizing these factors. The relative importance of each will vary depending upon the information retrieval system being employed, the storage
medium used, the speed of the search algorithm, the blocking factor of the
data base, the degree of overlap between I/O processing and CPU processing, and other factors.

A program has also been developed to merge the new SDF file into a
retrospective data base by division. This retrospective SDF file can then
be searched using the previously described SDF search system.

VI. INTERACTIVE SEARCHING

A project was begun to develop a user-oriented interactive search
system which would allow for a great deal of flexibility in the type of data
that could be searched, and would also be simple to use. It was decided to
develop an interactive system which could take advantage of the IBM 2741
typewriters and the cathode-ray terminals which are part of the University
of Pittsburgh Time-Sharing system. This approach seemed to eliminate
three disadvantages of the batch-oriented system. First, the batch-oriented

TABLE 5

Comparison of Retrospective Search Systems

TEXT-PAC Retrospective

Total number of citations

Number of profiles	Number of terms	Average number of terms/profile	Number of alerts	Average number of alerts/profile	Execution time	CPU time	Total cost($)	Cost/profile($)
1	16	16	13	13	11.45	1.97	28.64	28.64
8	189	24	676	85	16.07	12.88	40.18	5.02
12	272	23	609	51	23.31	14.56	58.27	4.86
20[a]	278	14	1738	87	34.29	22.93	90.76	4.03

Retrospective Search Using CASCON Programs

Total number of citations

Number of profiles	Number of terms	Average number of terms/profile	Number of alerts	Average number of alerts/profile	Execution time	CPU time	Total cost($)	Cost/profile ($)
13	181	15	1667	128	38.00	28.64	95.01	7.31

Retrospective Search of Chemical Titles Using Chemical Abstracts Service Programs

Total number of citations

Number of profiles	Number of terms	Average number of terms/profile	Number of alerts	Average number of alerts/profile	Execution time	CPU time	Total cost($)	Cost/profile($)
20	150	7.5	724	362	37.27	25.27	90.04	4.51

[a] A statistical report was requested on this run which would increase the processing costs due to punched card output.

TABLE 6

Retrospective Test Runs via HASP-II[a]

Number of questions	Number of terms	Elapsed execution time	CPU time	Number of alerts
1	8	1.99	1.06	58
1	14	12.40	1.69	58
1	30	12.19	2.22	58
1	58	8.20	2.86	58
1	117	13.14	3.29	58

[a]Physical and analytical division, Volume 70, covering six months of current-awareness data.

TABLE 7

Retrospective Test Runs via HASP-II[a]

Number of questions	Number of terms	Elapsed execution time	CPU time	Number of alerts
1	8	7.68	1.07	58
1	8	15.54	2.42	162

[a]Physical and analytical division, Volume 70, covering six months of current-awareness data.

search systems require that either the user have some knowledge of computers or that an information specialist be available to translate user search requests into meaningful search queries. Second, the majority of batch-oriented search systems seem to be limited to searching only a small number of data bases. This allowed the potential user very little flexibility in the types of data he might use as a data base. In short, most search systems were designed to process a specific file. Finally, a batch search system does not allow the user to monitor his output as the search takes place. The logic of the interactive search program was adopted from the Standard Distribution Format search system discussed in Section IV of this chapter.

In most cases, a user interested in building an interactive search data base would build the data base by reading punched cards onto a disk which is part of the Pitt Time-Sharing System (PTSS). The interactive search system has been built to accept free form input from punched cards. The advantage of this from the user's point of view is that a data base can be built of prose, lists of information, or any combination thereof. There is no rigid formating required of the input data, and consequently the user can search virtually any data base. However, several other features were built into the interactive system to make it more usable in a time-sharing environment.

Document searching with the PTSS interactive search program involves four phases: Pre-Input, Profile Input, Search, and Post-Search. Prior to entering the profile, the user may control the search with two options and the printing of the citations retrieved with one option. Typing $SEARCH=nnn will limit comparison between the profile and the data base documents to only the first nnn characters of each citation. Entering the phrase $STOPAFT=n will temporarily suspend searching after retrieval of n documents, allowing the user to check the effectiveness of his profile logic without searching the complete data base. Printing of the citations retrieved, which occurs in the Post-Search phase, can be controlled in the Pre-Input phase with the $LIST option. $LIST FIRST will selectively display only the first line (80 or less characters) of each document; and, conversely, $LIST ALL will display the entire document.

The second phase, Profile Input, supports only one option. Any unnecessary searches, caused by improper profile logic or misspelling, can be printed with the $CANCEL option. After typing this option the search program will disregard all previous profile input, return to the Pre-Input phase, and respond with "READY."

The third phase is initiated by entering the word "END" following the profile. Upon completion of the search, the program will indicate the number of references retrieved and ask the user if the citations are to be displayed. Answering "YES" will cause all citations to be displayed.

To facilitate the output of retrieved documents, the search program supports four options during the Post-Search phase. If the number of documents retrieved is large, or if the user is at a CRT and desires a hard copy of the retrieved citations, he may enter $PRINT, which will output the

citations to a high speed printer. As an alternative, $PUNCH will output
the citations to the system card punch.

If the user answered "NO" when asked if the retrieved documents are
to be displayed, or had specified $LIST FIRST before searching and would
rather have the entire document displayed, he may enter $DISPLAY, which
will redisplay the retrieved documents. Any of these three options,
$PRINT, $PUNCH, AND $DISPLAY, may be entered multiple times to
produce additional copies of the citations retrieved.

Two additional options are available during the Post-Search phase to
reduce the time required to conduct multiple searches on a large data base.
The user may enter $CATALOG, to create a new searchable data base
containing the documents retrieved by the last profile. Entering $EOJ will
then cause the search program to request the name of a new search file.
Using these two options, a user can quickly create a system of specialized
and efficient subfiles.

Five additional special-purpose options are also available. A search
which has been temporarily suspended through the use of $STOPAFT=n
may be continued by entering $RESUME. $COUNT will return to the user
the number of documents contained in the current search data base. $TIME,
entered during the Pre-Input phase, will display, upon completion of the
Search phase, the CPU time used for each search.

Four of the 15 available options, $STOPAFT, $SEARCH, $LIST, and
$TIME, require operands which may take on multiple values. $STATUS
will display the last operands specified for each of these options. The remaining option, $STOP, terminates execution of the search program.

Table 8 depicts sample timings for two interactive data bases of 5,000
and 50,000 card images. These tables indicate both the total elapsed time
(response time) and the central processor time (CPU time) required for
each search. While the search is conducted in a matter of seconds, the
total response time is measured in minutes. This is a function of the
Time-Sharing System. The Time-Sharing System must service all other
user programs as well as the search program. In many cases, this could
amount to 30-40 other programs. Consequently, response time is much
greater than search time.

An interactive search system servicing a heterogeneous population,
for example, a University community, must satisfy two criteria: 1) it must
be easy to use, and 2) it must provide all file generation and document dissemination services required with minimal or no user programming.

The Pitt Time-Sharing interactive retrieval system has met these
criteria through the use of free form text documents and profile input and
through the support of user options. The fifteen options currently supported

TABLE 8

Sample Timings for PCIC Interactive Search System

Chemical Titles File: 50,000 Card Images

Number of terms	Response time (min)	CPU time (sec)	Documents retrieved
2	7.0	12.74	0
4	8.5	20.10	88
6	10.5	24.52	86
10	18.5	28.15	86
12	31.2	30.17	92
14	16.1	26.12	100
20	27.4	29.30	315
24	36.5	34.61	298

Retrospective File: 5,000 Card Images

Number of terms	Response time (min.)	CPU time (sec)	Documents retrieved
2	2.6	1.52	59
4	4.5	5.08	65
6	4.7	5.11	66
8	8.3	8.48	66
10	8.5	10.36	91
12	10.0	13.83	181
14	10.3	16.66	175
20	14.6	22.89	202
24	14.5	22.60	256

by the system have eliminated all additional programming by the users.
The response from the users of the system has been very encouraging, and
it is expected that any new interactive retrieval system will provide a similar service. Additional information on interactive systems is given in
Chapter 5.

VII. CONCLUDING REMARKS

Concluding remarks have been written to provide the reader with some
indication of the considerations in organizing an information retrieval center. One of the first such considerations is the choice of a data base. Obviously, the data base must be germane to the user. The data base which
is appropriate for industrial chemists might not be the best choice for academic users. The potential subscriber should examine the data elements
in the file. For example, the presence of key words, registry numbers,
and molecular formulas should be noted. The journal coverage of the data
base should also be carefully reviewed. Royalty arrangements with the
tape supplier and computer tape capability between installations must also
be checked.

The next step is to decide whether to acquire a prewritten search system or to develop a tailor-made system. If an organization plans to implement a preprogrammed software package, the following factors must be
considered. The programs must be extracted from a source tape and assembled to produce an executable program. These programs in turn must
be fully tested to make sure that they are operational as a system. The
successful completion depends heavily upon the completeness of the documentation which was provided with the system as well as upon the ability
of the recipient of the system to communicate with the supplier. Other
technical problems could arise from differences in computer configurations
and operating systems. It is also difficult to make program changes to
programs which have not been written in-house. Finally, after the system
becomes operational, the output seldom proves to have all the features
which the users might prefer.

If a potential user decides to write his own retrieval system, another
set of considerations arises. One of the most important is to be able to
provide programming personnel with a full, detailed description of the
kinds of outputs required from the retrieval system and the kind of logic
capabilities which should be built into this system. This in turn requires
an in-depth knowledge of other retrieval systems and user needs. Skimp-
ing in the hiring of high-caliber data processing people only leads to the
development of worthless systems.

Programmer and computer costs will probably exceed those for a prewritten system. Whichever of these two approaches is taken, it is

imperative that user needs such as search logic capabilities and output options be taken into consideration prior to implementation. Incorporating user features after the system has been installed is often difficult and time-consuming. Another factor in the evaluation of a current-awareness system is the ease of adapting the system for retrospective searching. A system which performs a linear search can be used for retrospective searching. However, this "brute force" approach would probably be less efficient than a system specifically designed for retrospective searching.

Retrospective searching creates problems inherent with the manipulation of large files. In constructing a retrospective file, a decision must be made whether or not to include all the data elements in the current-awareness file. These elements (for example, corporate authors and addresses) need not be passed along in building the retrospective file. If possible, it is advantageous to subdivide the current-awareness file. For example, it was found useful to subdivide the Condensates file into five major divisions (i.e., five separate retrospective data bases).

Retrospective software considerations are heavily dependent upon the type of file organization. A brute force approach is easiest to implement, in that the current-awareness system is merely used to search multiple tape files. Of the various possibilities, however, this method is probably the least efficient. Any file reorganization will involve a trade-off between additional computer costs and improved search efficiency. In the limiting case of complete file inversion, search times are relatively rapid at the expense of file inversion costs. The number of inversions (updates) made to the file per year and the number of searches made against the file affect the feasibility of this approach. An associated limitation in developing an inverted retrospective file is the amount of random access storage required.

Early experiences in searching CT interactively did not prove to be favorable. Subsequently, an interactive search system that holds greater promise was developed. This system, employing a partially inverted file organization, is proving useful in searching a number of data bases of up to 50,000 card images. Caution should be used in searching significantly larger data bases interactively. Costs can mount rapidly due to file inversion, updating, and disk storage costs.

A number of factors contribute to information retrieval costs. Information retrieval requires large computers, thereby automatically incurring large costs. Search-processing costs are a function of a number of variables: input/output processing speeds, length of the data base, efficiency of the search algorithm, number of terms in the search profile, number of alerts retrieved, set-up times, data base conversion costs, etc. In addition, several man-years of effort can be expended in the programming and implementation of an efficient retrieval system. Another significant and sometimes hidden cost is in the area of support personnel required to maintain a large retrieval service. Information specialists trained in chemistry are required to translate the users' search needs into computer-readable search profiles. Abstractors, technical librarians, and office personnel are required. All of the above factors contribute to making computerized information retrieval expensive. The nature of the computing environment plays an important role in cost considerations. It is certainly desirable to have a dedicated computer to speed systems development. An over-the-counter environment increases implementation time and hinders user service. Each retrieval center must address the question of how large a user community is needed to overcome these costs. At the same time, there is a limit to the amount a user of these services is willing to pay. Providing a reliable current-awareness and retrospective service will require adequate resources, careful planning, and talented programming support.

REFERENCES

1. E. M. Arnett, "A Chemical Information Test Station," Chemistry, 42, 16 (1969).
2. E. M. Arnett, "Computer Based Chemical Information Service," Science, 170, 1370 (1970).
3. R. Freeman, J. Godfrey, R. Maizell, R. Rice, and W. Shepard, "Automatic Preparation of Selected Title Lists for Current Awareness Services and Annual Summaries," J. Chem. Doc., 4, 107 (1964).
4. Data Content Specifications for CA Condensates in Standard Distribution Format, Chemical Abstracts Service, Columbus, Ohio, 18, 1970.
5. Standard Distribution Format Technical Specifications, Chemical Abstracts Service, Columbus, Ohio, 1970.
6. G. Badger, E. Johnson, and R. Philips, "The Pitt Time-Sharing System for the IBM System 360: Two years Experience," Proceedings of the Fall Joint Computer Conference, San Francisco, December 1968.
7. "Pitt Time-Sharing System for IBM System/360," Computer Center, University of Pittsburgh, Revised March 1968.
8. M. J. Bloemeke, and S. Treu, "Searching Chemical Titles in the Pittsburgh Time-Sharing System," J. Chem. Doc., 9, 155 (1969).
9. A. V. Esposito, R. Fleischer, S. D. Friedman, S. Kaufman, S. Rogers, S. Skye, and M. Shotkin, "TEXT-PAC, S/360 Normal Text Information Processing, Retrieval and Current Information Selection Systems 360d-06.7.020," IBM, Armonk, New York, December 1968.
10. N. Grunstra and K. Jeffrey Johnson, "Implementation and Evaluation of Two Computerized Information Retrieval Systems at the University of Pittsburgh," J. Chem. Doc., 10, 272 (1970).

11. E. Howie and A. Kent, "SDF Search System Usage Manual," The Knowledge Availability, Systems Center, University of Pittsburgh, Pittsburgh, Pennsylvania.
12. E. Howie and A. Kent, "SDF Search System Implementation and Maintenance Manual," The Knowledge Availability Systems Center, University of Pittsburgh, Pittsburgh, Pennsylvania.

Chapter 5

INTERACTIVE RETRIEVAL SYSTEMS

Elaine Caruso

Pittsburgh Chemical Information Center
Interdisciplinary Doctoral Program
in Information Science
University of Pittsburgh
Pittsburgh, Pennsylvania

I. INTRODUCTION

Interactive retrieval systems require fast access to large stores of data. Since creation and storage of the programming systems and data bases is expensive, they should be useful and accessible to large numbers of people who can share the cost of the service they provide. To maximize accessibility the service should be available many hours during the day, and entry points to the system should be available at places where potential users are located. Direct access storage devices, available since 1949, provide required access speeds and storage capacities; time-sharing systems and communication with computers provide the desired temporal and geographic entry spread, so that a user may reach such a system at any time, from any place which can be reached by phone. Large stores of machine-readable data, for the most part incidental by-products of computer-aided production of printed indexing and abstracting publications, are also available.

The design and development of interactive on-line search capabilities has been the subject of much thinking and writing during the few years that the required technology has been a reality. A quick check of the various indexes to the five issues of the Annual Review of Information Science and Technology [1] provides us with 31 references. More than 30 such systems* are known to this author, from the document and report literature. Yet it is doubtful if there exist reports of controlled use of these systems by more than 100 individuals. Most of the literature on interactive systems is predictive or descriptive; very little use experience can be gleaned. Reports which do recount attempts to assess, however subjectively, user search activities include: Freeman and Atherton's AUDACIOUS experiment (1968) lists "perhaps 50 (participants) as users or observers [2];" Meister and Sullivan's evaluation of RECON (1967) [3], 38 users; and Caruso's symbolic strategy tutorial system (1969) [4], 35 users. The latest activity report of Project Intrex (March 1970) [5], reports three experimental users of the interactive search system during a six-month period.

Even though user experiences were recorded in "controlled" situations in these studies, only the INTREX studies required that participants have real need for the information sought via the systems. As outlined in the step-by-step experimental procedure for user interaction at INTREX, established under the direction of Professor J. F. Reintjes, the first condition requires that "the user ... have a real need for information in ... the areas for which our data base is designed." No argument can be made about the necessity for this constraint on user studies. "Within this restriction we want to have a wide range of user types ..." concludes the Reintjes procedure [6] for choice of users.

There is some need, at this point, to investigate reasons for the scant experimental use records, and many suggest themselves. The suitability of incorporated data bases to the interests of the available user populations is probably the single most critical factor limiting our accumulation of

---

*A survey of reported on-line reference retrieval systems, with evaluation of the most distinguishing characteristics of each, can be found in Chapter 2 of Ref. 4.

evidence about "real" use of any mechanized retrieval system. While the ubiquity of the telephone suggests that an appropriate user-group should be readily available for any given data base, its use to extend geographic access to that data base is not economically feasible. Some less expensive communications channel is needed to expand user groups to sizes large enough to result in more economical higher-use rates, and provide candidates for experimental manipulation in controlled tests.

We have an alternative to the possibility of extending the geographic range of our system to increase the size of our user group; we can incorporate into our system not one but many of the available machine-readable by-products of computer composed indexes and abstract publications. However, since these available data bases require extensive reformating and reorganization to allow the fast access essential to on-line interaction, and the accordingly complex updating programs, again we are limited by cost considerations.

Obviously, until we can adjust various problems to allow us to develop large user-groups for interactive search services, we must make every attempt to evaluate, in its own context, all available user-interactive system experience.

We have stated the need for sharing results of any user-system engagements. We further state that, given the wide variations in systems, data bases, and experiential environments, and in particular, the impossibility of separating the interaction of the experimenter with the user from the user-system interaction, a case study type of reporting can provide much benefit that might be obscured in fragmentary reporting of the results of carefully "controlled" testing of systems. Further, it can provide in an economical, readable way the answers to the myriad questions being asked of interactive systems [7], and preserving information which may be unquestioned as yet.

As anyone who has ever attempted to introduce a novice to any use of a time-sharing system can testify, it is virtually impossible to anticipate and make formal provision for all the variations of user/system difficulties which may occur. In conducting the experiment in use of the symbolic-strategy tutorial, the author was unusually fortunate, experimental use being interrupted only twice in 37 attempts by system failure, and having to intervene to interpret system instructions only in five cases. The short time span of the experiment (one week) explains this "luck" in part, together with the obduracy of the experimenter who made it clear to the subjects that they were expected to find their own way out of any difficulties [8]. This problem is attested by Morrill in describing his effective training program for a computer based management-information system as being too costly to mechanize and better handled by a human aide [9]. The IN-TREX experimental procedure referred to earlier, recognizes the

inevitability of experimenter intervention and specifically details the situations, "briefing the user," urging that the user proceed unaided but allowing the experimenter to help him if he has difficulty. A tape recorder
is used to record these interactions which cannot be detected and automatically recorded by the programmed system.

II. INTERACTIVE RETRIEVAL PROGRAMS DEVELOPED AT CIC

A. Circumstances

When the ideas to be incorporated into the chemical information experiment were brought together early in 1967, a tutorial program was in the
final stages of development. This program included an interactive on-line
search of an indexed-sequential data base. Since our early identification
of experimental objectives included an effort to develop new ways to introduce and use computer-based information services, it seemed natural and
obvious that interactive retrieval be included as a major area for study.

Two different on-line projects within the Chemical Information Center
were initiated. One, involving the dominant direction of CIC progress, was
an attempt to use on-line searches to prepare output for a weekly selective
dissemination of alerts culled from Chemical Titles. The quick disillusion-
ment which resulted is reported in "Searching Chemical Titles in the Pittsburgh Time-Sharing System"[10]. A second task group was assigned the
objective of developing interactive applications. It is with this "group" that
we are concerned in the remainder of this chapter.

B. The Environment

We will not attempt a comprehensive description of computer facilities
since they are well documented in Chapter 2. Special equipment available
to the Interactive Applications Task Group (IATG) initially included only the
IBM 2741 typewriter terminal; later the hard-wired 720 Display Scope became available, and more recently the portable Datapoint 3000 was acquired.
The only programming language with extensive string-processing and file-
manipulation capabilities available on the time-sharing system is PIL, an
interpretive JOSS-like language. With its interruptible execution, built-in
editing capabilities, and ease of revision, PIL is an ideal language for the
development of experimental interactive programs; successive revisions
of PIL have made it economical to run PIL tutorial and search programs
using small files routinely [11].

The potential user population included students, faculty, and research
associates of the Chemistry Department of the University of Pittsburgh

and of the Mellon Institute, and chemists employed by industry in the Pittsburgh area; essentially the same user group to be served in the Chemical Information Center's selective dissemination project. Our user-group choices were severely restricted. As it turned out, the user-group became the students in a class on chemical literature, plus a few volunteers from one of the last of the seminars held to introduce CIC services to new users.

Four data bases on magnetic tape were chosen as target files for interactive exploitation by the intended, appropriate user-groups. (For much more detailed description of these programs, see reference 12):

1. The Chemical Abstracts Condensates - citations and keywords issued weekly by Chemical Abstracts Service.

2. The Sadtler Corp. Infrared Spectral Data - the data, with microfilm copy of source documents.

3. The Nuclear Magnetic Resonance bibliographic file augmented with indexing terms (Preston Laboratories).

4. Crystallography Data Files, produced by the Crystallography Department and the Knowledge Availability Systems Center of the University.

C. The Crystallography Data Files

The Crystallography Data Files were quickly rejected as candidates for interactive search programs; while access via the remote terminal was desirable, the already available programs, formerly used in off-line, batch mode on the IBM 7090, were entirely adequate and were adapted for use on the time-sharing system. The compactness of the coded data and the small size of the file made no excessive demands on the disk storage available.

D. The CAS Condensates Data Base

It was obvious that any attempt to provide interactive search on any current basis for the weekly Condensates issues was impossible with available resources. However, the programming staff of CIC was considering building an indexed sequential file of the Condensates issues as they were searched for weekly SDI's (an optional capability of the IBM TEXT-PAC search program). We determined to adopt a three-phased program utilizing the Condensates service as the source of our data base:

1)   Development of a tutorial program, with the objective that the user should be able to more effectively write a Condensates-type profile using the conjunction of disjunctive parameters, negation, and truncation features only.

2) Development of the practice search capability of the Phase I
tutorial and increment of the data base to the point where users could fully
develop profile statements for the SDI service on-line, the statements to
be automatically entered into the TEXT-PAC jobstream at stated intervals.

3) Recording of the PIL-programmed interactive search algorithms
into assembly language for efficient searches on the huge retrospective
data base when it became available.

Active work on the Condensates files began early in 1968. The first
tutorial was patterned after the general-strategy tutorial [13], which took
a hypothetical question appropriate to the data base, developing it from a
simple single aspect generic search into a fully refined specific question.
At each stage the user was required to specify search terms and their
logical relationships. The use of left- and right-truncation and negation
were prompted. User terms and strategy inputs resulted in actual searches
with results printed out in a modified version of the output of a TEXT-PAC
search. Following one "guided" run of the search program the user was
urged to experiment with other questions of his own.

The original strategy tutorial had required about two and one-half hours
for an average user to complete. In order to reduce user time at the console,
we eliminated the device of requiring the user to read an accompanying
text at intervals before allowing him to continue. Instead, a series of
"Helps" was stored, which could be called whenever the meaning of a sequence or an interpretation of its effects were not clear. Printed text to
accompany the program was reduced to a minimal single page which identified the program and its purpose and enabled the user to turn the terminal
on, load the program, and to restart the program in case some error condition should cause the program to stop execution.

Preliminary use tests by 12 graduate students in the Department of
Chemistry gave disappointing results; unlike the users of the earlier, more
didactic tutorial program, they did not grasp the use of synonyms in OR'd
parameters nor the effect of AND'd parameters. The use of truncation to
make matches with word stems or suffixes was grasped. But since the
data base consisted of uncontrolled vocabulary, effective searches could
not be made unless the user himself supplied all possible synonyms for the
desired concepts. The use of logical conjunction (AND'd parameters), to
reduce query output to specifically wanted topics, where several more
general concepts are wanted only when they relate to other more general
concepts, is also essential when searchable keys are not "role-linked" or
precoordinated.

Even more disappointing is the fact that the carefully composed and
paced "Helps" were never called out. Occasionally the system prompted
their use, when a mistake occurred repeatedly in the same place in the
program, but the students never took the initiative of ignoring the program's

current "prompt," to type instead the word "Help" as they were told to do in the first few lines of instructive output. Whether they forgot, or whether they lacked the courage to depart from the pattern of prompted input, was not determined; they did comment that they never really understood what the were supposed to do, however.

Meanwhile, the Pittsburgh Time-Sharing System had added CATALYST, a CAI (Computer-Assisted-Instruction) language to its repertoire, so we wrote a short instructional sequence of the programmed instruction, frame-by-frame type to clarify the use of logical sum, product, difference, and negation. Pseudopractice* in their use is required. This sequence precedes the training search. We reduced tutorial elements of the search, retaining the "Helps," however. Student users do show more understanding in the use of logical connectors and are less confused in program operation. Most complain that the one-hour session is too long.

E. Sadtler Corporation Infrared Data Files

This file is derived from the infrared spectra of 29,000 organic compounds. Infrared spectra represent, in effect, "fingerprints" of different types of molecules, although they are to some extent interpretable in terms of molecular structure. The data are presented as a continuous, linear portrayal of peaks and valleys representing the intensity of infrared absorption by the compound at different wavelengths (expressed in microns) of infrared light.

These files, though highly structured and consisting primarily of coded data, were otherwise very suitable for interactive searching. The user approaches the file most often with his own interpretation of a spectrum and with a recogniztion of his own uncertainty of the correctness of his search key. The system could, with speed and accuracy, report the results of attempts to match several variations on the search prescription. The user could be expected to recognize the correct answer to his query with a fair degree of confidence. The data base, relatively small and up-dated annually, was one that our small staff could hope to keep current.

---

*The CATALYST language did not permit incorporation of files on which effects of logical strategy variations could be tested. By careful question framing, and the use of a term-document matrix, we were able to simulate the effects of various Boolean prescriptions as if searched on a given matrix of terms/documents. An illustration of the use of this program and a listing of the program are included as Appendices II and III of the Second Annual Report of the Pittsburgh Chemical Information Center [14].

The program, as developed for the spectral data, searched only two categories of the available data on-line; the location of strongest peaks in each micron area and the chemical "class" of each recorded compound. Results of the on-line search gave the user the serial number of the microfilm record which contained the full spectral data for the identified compound. If the Chemistry Library terminal were used, he had only to hand the output to the librarian to be given appropriate photocopy.

If searches on other data elements were desired, the on-line program accepted the request, coded and formatted it as required by the Sadtler COBOL search program, and added it to a query, with appropriate job control statements, to be run in batches at stated intervals. The program was written, debugged, tested; in short, brought to operational status. No users appeared. Perhaps this was due to a failure to promote the service, since the Chemistry Department is housed some distance from the offices where the program was developed. There was no contact with students or faculty. The librarian simply reported no call for the service.

Recently Sadtler Corporation demonstrated its own new on-line search capability, InfraRed Information System (IRIS), for the students and faculty of the subscribing organizations. While quite fast, turnaround being a matter of minutes, the search is not interactive. The user does have one search option which he can vary. He can ask that his search key be "wiggled" or searched in micron areas adjacent to that specified. Results may be printed as soon as completed, or he may have them stored, to be printed later on request.

F. NMR Bibliographic Files

The Preston files on references to nuclear magnetic resonance data were available in the Chemistry Department as a series of about 20,000 edge-notched cards, searchable on coded keys only. Users had been reported to attempt to locate needed references not just by "needling" their way through box after box of cards, but just thumbing through the cards reading them one by one.

The Preston tape and their search program for the off-line batched searches were acquired, and the program was brought to operational status on our batch-mode IBM 360/50. An on-line program was planned to accept user queries, and code and format them for the off-line search. It had become fairly evident in the meantime, however, that user demand for such service did not exist; or if it did, it was too infrequent to justify costs of maintaining the service on-line.

G. Summary

Our efforts at interactive systems developments have resulted in the
development of several programs: a tutorial for preparing Condensates-
type profiles and a search program for small disk-stored files; a program
which searches infrared data to match peaks in given micron areas over a
portion of a file, stores the results and the queries for batched searches
of the entire file off-line; preliminary searches for bibliographic sources
of NMR files. No real-life use has occurred, partly perhaps because busy
people, while interested, will not take time to learn a new, only indirectly
useful skill. In some instances (Condensates, Preston NMR), the data
bases never reached useful sizes; and in others (Infrared, NMR) the use
rate is naturally quite low.

One last item must be added to this chapter. Early in 1969 one of the
programmers working on the adaptation and implementation of TEXT-PAC
sat down to the on-line terminal and played the role of user of the symbolic
strategy tutorial program. Then, in odd moments, he sat down and wrote
an extremely simplified version of an interactive search using AND and OR
logic. In addition he wrote a file-generating program which took any punched
or tape-stored file and made it suitable for use with the search program
[15].

Response from within the University was immediate; individual users
have appeared in the Chemistry Department, the School of Engineering,
the Hillman Library, the graduate schools of Public and International
Affairs and of Library Science; probably others use the system, but we
currently have no way to determine this. A recent discussion of the system, dubbed SEARCH, drew an audience of 50 members of the Association
of University Librarians, on their lunch hour! The current SEARCH
Manual [16] lists 22 data bases which their creators are willing to share
(see Fig. 1)

Why was the response to and use of this program so markedly different from response to earlier systems ? For one thing it is easy to use.
One command brings it on line; a second loads a data base to be searched.
The search statement can be as short as two lines, two words, of input.
Pittsburgh Time-Sharing System itself has become more reliable, easier
to use. The sophistication of the user-community has increased tremend-
ously.

These reasons all contribute to the immediate acceptance of SEARCH,
but the one characteristic which is most critical is that cited earlier, the
suitability and value of the incorporated data base to the potential user
community. A user's own files, easily punched in any format, or using
previously punched or recorded files together with other user's files of
highly selected materials, provide a data base, indefinitely extensible, of

1. AUTFIL:SEA(Q54dec)
   Approximately 70 documents on LIBRARY AUTOMATION

2. AUTLIB:SEA(Q54dec)
   Documents on INTERACTIVE SYSTEMS, CAI, and LIBRARY AUTO-
   MATION in QUIC format

3. MARC2:SEA(Q54dec)
   Approximately 392 documents in MARC II format from the MARC II
   test tape produced by the Library of Congress

4. MARCENG:SEA(Q54dec)
   Approximately 180 documents in MARC II format produced by the
   Engineering Library

5. SPIN1:SEA(Q54dec)
   Documents on NUCLEAR PHYSICS from the SPIN/O tape produced by
   the American Institute of Physics

6. SPIN2:SEA(Q54dec)
   Documents on PHYSICS from the SPIN/O tape produced by the
   American Institute of Physics

7. UPCHEM:SEA(o82dec)
   Contains 58 documents on POLYMERS

8. AIPPIL:SEA(Q54jgw)
   Approximately 100 documents on URBAN AFFAIRS in QUIC format

9. JON:SEA(Y92ncc)
   List of theses produced by GSPIA students

10. CA66:SEA(Q54dec)
    Approximately 100 documents on SYNTHETIC HIGH POLYMERS
    from Chemical Abstracts, Vol. 66, 1967

11. MARCTINY:SEA(Q54dec)
    Approximately 50 documents in MARC II format

FIG. 1.   Examples of searchable datasets available.

the highest possible interest level. Complete privacy of use, as well as
free sharing of files, is possible.

On-line tutorials, as well as the printed manuals, are available both
for file generation and search processing. These have been undergoing
preliminary use trials. Problems in accommodating both the upper case only
Datapoint 3000 and the upper/lower case capabilities of the 2741 terminals
have been studied. Automatic recording capabilities to collect use statistics are being incorporated into both the tutorial programs and the search

programs. We have nearly ready a research instrument which we can be assured will be used in "real" information processing situations. We have already learned much in the development of this program, as the programmer, sensitive to freely expressed user comments, has developed and extended the capabilities of the program. We are eager, now that the development of the program has stabilized (to a degree), to begin to observe actual use patterns in real-use situations.

H. The Future

In "a possible future" [17] B. C. Vickery proposes an inversion of the memex [18] concept as the end product of an extrapolation of current trends in information retrieval systems. The scholar's desktop library of handbooks, current journals, and photocopied articles will continue, but augmented by console access to machine-searchable stores of document references, full text, data, subject maps, and lists of individuals interested in particular topics. He will have access to increasingly larger stores as he moves from local to national stores, from current alerts to digests and reviews of his field to retrospective searches of centrally maintained cumulations of indexes and abstracts. Using the same console that makes available the calculating power of the computer, he may conduct "a dialogue with the information store (which) may in itself be the most effective way of educating the scientist in the use of information sources ..., offering the further possibility of making contact with other users with similar interests ... in short, ... approaching more nearly the informal dialogue which is the ideal mode of communication" [19].

Without emphasis, Vickery has clearly set forth clearly the objective which should dominate the thinking of designers of interactive retrieval systems. These systems should be developed to allow an ultimate, informal, one-to-one type communication between individuals who generate and use information. Current efforts to develop analogs of the user-librarian interview [20] or the user-card catalog interaction [21] are far too limiting for effective use of presently available machine-readable stores and communication capabilities. In point of fact, all presently operational or actively developing systems for interactive search dialogs have a much more narrowly defined subject content that the card-catalog analogy would allow. They very closely resemble discipline-oriented subject indexes in content organization, user-population, and use. This is not by design, but expediency. The data bases to support the systems are by-products of printed index production processes: RECON [22] uses the National Aeronautics and Space Administrations's tapes spun off from printing of STAR; National Library of Medicine's Biomedical Communication Network will use tapes of Abridged Index Medicus [23]; Excerpta Medica's on-line search uses photocomposition computer tapes; Epilepsy

Abstracts, interactively searchable by Data (Central) programs, uses similar tapes [24]. Other sources of machine-readable data bases, the only
economical input for large store systems, betray similar origins. This
inadvertent mimicry is as limiting as the deliberately sought-after anal-
ogies to card catalog and reference librarian.

Let us see where the objective of informal communication between
individuals might lead us.

A recent workshop [25] brought together 40 individuals, all interested
in a relatively narrow area, the man-machine interface, within the more
general area of the design of interactive retrieval systems. While highly
selected for interest in and contributions to the topic, most not only did
not know each other, but had no familiarity with one another's reported
efforts. All had access to terminal devices for use in their work, and
probably "dial-up" devices with acoustic couplers for computer connections
via telephone lines. Perhaps some had bibliographies for their collection
of reprints, photocopies, and reports, in machine-readable form. Here
seems to be a natural beginning place for establishing a computer network,
via phone lines, for sharing resources. A simple Boolean search system,
based on full-text searching (such as SEARCH, which is described earlier
in this chapter) could be made available with each file. Informal notes,
progress reports, research efforts of students—all could be worked into
such a system. Provision could be made for some dialog between individuals quite easily. With a minimum of investment, we would have a pool
of the kind of "experienced users" of the type Douglas Engelbart has so
wisely suggested [26] is necessary before any serious comparison of variations of interface design can be made. Further, we might discover that
use of interactive searches is readily understood and eagerly adopted when
data bases are sufficiently interesting. Here we have a possible direction for
system development which bears little resemblance to that system, patterned after the "interactive" card catalog; one which might suggest alternatives to elaborate self-training systems, which could develop a "Highway
101" effect. For further motivation for this type of communication devel-
ment, see John B. Calhoun [27] on the enlargement of conceptual space, in
his AAAS address, December 1968.

Presently conceived information retrieval systems are based on the
large store myth; i.e., to be of interest the data base of the system must
encompass all of the literature on some definable subject area, several
hundred thousand documents, minimum. Is it not true that, for a given
user, only a small subset is wanted or can be used? Is it not therefore
true that, for a given user, the interest level of a stored file is directly
related to its smallness, where that small size is a reflection of the select-
ivity which was exercised in its creation? The only valid reason for creating huge stores is to allow the individual creation of a multitude of small
subsets of that store, i.e., answers to requests for documents on relatively

narrow subjects. A system which has as its nucleus no given data base, but rather a network of people communicating information, which may or may not be in any publishable form, may be the proper use of the computer information network.

REFERENCES

1. Annual Review of Information Science and Technology, Encyclopaedia Britannica, Chicago, Vol. 1-5, 1966-1970.
2. Robert R. Freeman and Pauline Atherton, "An Experiment With an On-line, Interactive Reference Retrieval System Using the Universal Decimal Classification System as the Index Language in the Field of Nuclear Science," (Report No. AIP/UDC-7), American Institute of Physics, UDC Project, New York, April 1967, p. 26.
3. David Meister and Dennis J. Sullivan,"Evaluation of User Reactions to a Prototype On-line Information Retrieval System," NASA CR-918, Bunker-Ramo Corp., Canoga Park, California, October 1967, p. 11.
4. Dorothy Elaine Caruso, "An Experiment to Determine the Effectiveness of an Interactive Tutorial Program, Implemented on The Time Sharing IBM System 360, Model 50, in Teaching a Subject-Oriented User to Formulate Inquiry Statements to a Computerized On-line Information Retrieval System," Ph.D. Dissertation, University of Pittsburgh, 1969, p. 69.
5. Semiannual Activity Report, (September 15, 1969-March 15, 1970), Project Intrex, Massachusetts Institute of Technology, Cambridge, Mass. March 1970, p. 19.
6. Semiannual Activity Report, (September 15, 1969-March 15, 1970), Project Intrex, Massachusetts Institute of Technology, Cambridge, Mass. March 1970, p. 17.
7. John L. Bennett, "Interactive Bibliographic Search as a Challenge to Interface Design," Challenge paper prepared for "User Interface for Interactive Search of Bibliographic Data Bases" workshop, August 1970.
8. Dorothy Elaine Caruso, "An Experiment to Determine the Effectiveness of an Interactive Tutorial Program, Implemented on the Time Sharing IBM System 360, Model 50, in Teaching a Subject-Oriented User to Formulate Inquiry Statements to a Computerized On-line Information Retrieval System," Ph. D. Dissertation, University of Pittsburgh, 1969, p. 41.
9. Charles S. Morrill, "Computer-Aided Instruction as Part of a Management Information System," Human Factors, 9, June 1967, p. 253.

10. Mary Jane Bloemeke and Siegfried Treu, "Searching Chemical Titles in the Pittsburgh Time-Sharing System." J. Chem. Doc. 9(3) 1967, 155-157.

11. PIL/L: Pitt Interpretive Language, Computer Center, University of Pittsburgh, 1970.

12. Dorothy Elaine Caruso, "Tutorial Programs for Operation of On-line Retrieval Systems." J. Chem. Doc. 10(2), 1970, 98-105.

13. Dorothy Elaine Caruso, "An Experiment to Determine the Effectiveness of an Interactive Tutorial Program, Implemented on the Time Sharing IBM System 360, Model 50, in Teaching a Subject-Oriented User to Formulate Inquiry Statements to a Computerized On-Line Information Retrieval System," Ph. D. Dissertation, University of Pittsburgh, 1969, pp. 25-32.

14. Annual Report, Chemical Information Center, 1968, Appendix II, III.

15. Elaine Caruso and Anand K. Gupta, "A Users' Manual for SEARCH:PTSS File Construction and Search Programs," Pittsburgh, December 1970.

16. Elaine Caruso and Anand K. Gupta, "A Users' Manual for SEARCH: PTSS File Construction and Search Programs," December 1970, p. 11.

17. B. C. Vickery, Techniques of Information Retrieval, Archon Books, Hamden, Connecticut, 1970, pp. 205-208.

18. Vannevar Bush, "As We May Think," Atlantic Monthly, July 1945.

19. B. C. Vickery, Techniques of Information Retrieval, Archon Books, Hamden, Connecticut, 1970, p. 208.

20. Morris Rubinoff, Samuel Bergman, Winifred Franks, and Elayne R. Rubinoff, "Experimental Evaluation of Information Retrieval Through a Teletypewriter," Communications of the ACM, 11, 1968, 599.

21. Lawrence H. Berul, "Document Retrieval" in An. Rev. Inf. Sci. and Tech., v. 4, 1969, pp. 203-204.

22. What NASA/RECON Can Do For You, National Aeronautics and Space Adminstration, Washington, D. C., November 1970, p. 2.

23. Herbert R. Seiden, A Comparative Analysis of Interactive Storage and Retrieval Systems with Implication for BCN Design, System Development Corp., Santa Monica, California, January 1970.

24. J. F. Caponio, J. K. Penry, and D. J. Goode, "Epilepsy Abstracts: Its Role in Disseminating Scientific Information," Bull. Med. Lib. Assoc. 58(1), 1970, 37-43.

25. "The User Interface for Interactive Search of Bibliographic Data Bases," a workshop held at Palo Alto, California, January 1971.

26. Douglas Engelbart, Proceedings of the Workshop on the User Interface for Interactive Search of Bibliographic Data Bases, Palo Alto, California (January 1971), in press.

27. John B. Calhoun, "Space and the Strategy of Life," in The Use of Space by Animals and Man (A. H. Esser, ed.), New York, Plenum Press, 1971.

Chapter 6

APPROACHES TO THE ECONOMICAL RETROSPECTIVE MACHINE-SEARCHING OF THE CHEMICAL LITERATURE

Bahaa El-Hadidy and Daniel James Amick*

Pittsburgh Chemical Information Center
Knowledge Availability Systems Center
University of Pittsburgh
Pittsburgh, Pennsylvania

and

Pittsburgh Chemical Information Center
University of Pittsburgh
Pittsburgh, Pennsylvania

139

---

*Present address: Department of Sociology, University of Illinois at Chicago Circle, Chicago, Illinois.

I. INTRODUCTION

The research scientist is confronted with two problems concerning the literature—namely, obtaining the results of the previous research efforts in a given area (retrospective searching) and keeping up to date on current progress in his chosen field of specialization (current awareness). This report is devoted to the problem of retrospective searching in the chemical literature, i.e., searching past literature when entering a new field or when completing coverage of a more familiar subject.

Traditionally, scientists have used the secondary literature as a tool to gain access to primary scientific and technical literature. While it is possible to scan the primary literature for current awareness, it is necessary in a retrospective search to depend mainly on secondary services. This is due to the fact that the literature in its original form (which is arranged in purely chronological order in each of many periodicals) is unsuitable for the purpose of manual retrospective searching since information relating to a topic is scattered in an almost completely random manner. One must therefore refer to literature which has been reorganized according to topic, as is the case in secondary sources.

Traditional secondary services in science and technology vary greatly in scope, coverage, and growth. They range from abstracts to indexes, bibliographies, reviews and library catalogs. Because of their continuing bibliographic service, abstracting and indexing services are the principal means available for providing scientists and technologists with retrospective as well as current-awareness services. These services provide scientists with the intellectual analysis and organization of the scientific literature which allows the literature to be handled systematically and effectively.

In recent years, the continuing, virtually exponential growth of the primary literature has strained the capabilities of the traditional abstracting and indexing services for providing their users with appropriate, timely, and comprehensive information support. The consistent eight to nine percent annual growth in the number of primary papers abstracted in Chemical Abstracts (CA) has not only caused a tremendous growth in its volume, but has also increased the number of its different types of indexes. Table 1, which shows the growth in the Collective Indexes of CA, illustrates the magnitude of this problem.

In order to provide users with additional points of entry to this ever growing mass of complex information, CA has increased the number of different types of indexes over the past decade from seven to thirteen. This has posed some formidable problems, not only in terms of production, but also of economy, timeliness, and use. The size and complexity of such indexes made it quite cumbersome for scientists to use them effectively for answering retrospective questions.

TABLE 1[a]

CA Collective Indexes

Series	Years	Pages	Volumes
1	1907–16	4,823	4
2	1917–26	6,591	5
3	1927–36	9,304	6
4	1937–46	11,241	8
5	1947–56	21,926	20
6	1957–61	22,864	15
7	1962–66	41,626	24
(8)	(1967–71)	(74,300)	(41)
(9)	(1972–76)	(110,700)	(62)

[a] Taken from Ref. 1.

About a decade ago, when it became evident that the traditional manual system for processing, publishing, and using the secondary literature was too slow, too expensive, too rigid, and too wasteful, Chemical Abstracts Service began shifting to a computer-based processing system. Beginning with 1961, CAS began producing Chemical Titles on magnetic tapes in addition to the traditional hard copy. Since then a number of other computer-based search services have been developed, which include Chemical-Biological Activities (CBAC), Polymer Science & Technology (POST), CA Condensates, and recently the Volume Indexes to Chemical Abstracts. CAS's basic target is a system in which all information will be input to one unified store, and from which a range of general or more specialized services can be supplied.

The use of computer and data processing equipment to process and repackage information may well be the most important development of the decade for affecting the capability of secondary services to meet user requirements. The computer's ability to read the store of material rapidly, compare the contents of an inquiry, and select from the store those items which satisfy the inquiry, greatly exceeds manual capabilities and renders computer-based systems especially suitable for retrospective searching. However, much experimentation and experience is needed before such a service can be used effectively and economically. The main problem in

developing such a service is the extremely high computer costs associated
with processing and using the very large file.

One of the basic aims of the Pittsburgh Chemical Information Center
(PCIC) has been the development of an efficient and low-cost retrospective
search capability for CA Condensates. With the successful development
of the current-awareness operation for CA Condensates at the Center using
the semi-inverted file TEXT-PAC system, there was no fundamental reason
why the same basic set of systems could not be used for retrospective
searching of the same tapes. During the year 1971 our basic efforts were
devoted to such operations. In the following parts of this report, the efforts involved in developing the retrospective system and the experience
gained from operating the system are described. The study of user information needs and the degree to which the retrospective service satisfied
these needs are also discussed.

II. DEVELOPMENT OF THE RETROSPECTIVE FILE

In developing the Condensates retrospective search system, the two
main criteria which have been taken into consideration by the PCIC are 1)
the computer cost, which is the biggest single cost of the system, should
be reasonably economical so that a relatively low charge for the service
would be possible and 2) the system should be designed to answer search
inquiries effectively without burdening the user with a large amount of
unwarranted material.

The development of such a system depends on two main elements:
first, a set of computer programs that will reorganize and assemble the
computer-readable data into a form where large searches can be done
relatively cheaply (i.e., shorten computer time); and second, a file organization capable of producing files of manageable size.

The current-awareness tapes, which are assembled to form the retrospective file, are serially organized so that one item follows another in
the order of their original accession and the whole must be scanned to locate a single reference. Such organization may be viable for current-
awareness searches since it is possible to batch some 200 questions prior
to the search, and each citation on the tape is examined only once. However, in retrospective searching, batch processing of questions is not
always possible, and the size of the store of information which accumulates
through a tape-oriented SDI system becomes enormous. Therefore, some
other ways of organizing the file had to be explored.

It was originally thought that a completely inverted file, in which each
attribute (term, concept, descriptor, etc.) is followed by all item numbers,
would allow great economy for retrospective searching. However, the cost
of the inversion operation and file update, as well as the considerable

investment in computer software, made this proposition unacceptable. As an alternative, it was decided to use the IBM TEXT-PAC system.

The basic advantage in using TEXT-PAC for retrospective searching lies in its unique file organization. The individual current tapes are processed by a special text conversion program. This conversion operates on the data, record by record, and arranges it alphabetically by word and word length, and carries with it associated information required to satisfy searching. That is, the system takes into account the position of the word in the record, in the sentence, in the author or title or keywords, its print fields, upper/lower case indicators and the subsequent alphabetical word regardless of word length.

Major problems in developing the Condensates retrospective search system have been the extremely large size of the file and the rapid rate at which it grows. For example, in the period between July, 1968 and December, 1970, the five volumes of Condensates available in machine-processable form contained 642,314 citations. Each week the file accumulates an average of 6000 records amounting to over 300,000 records per year. Assuming that each record contains about 25 words, and that the average number of characters per word is 6.7 [2], we have been faced with a file which has more than 100 million characters, and which accumulates about 48 million characters each year. It is clear that searching such a file is economically unfeasible.

To reduce the cost of searching the file, some criteria for dividing the data base into subfiles of manageable length were sought. One possibility was to subdivide the file solely by year, i.e., chronologically, which would require users to limit their searches to the most recent tapes. Some information retrieval systems, such as NASA's, follow this arrangement. However, as will be seen in the following sections, our user studies showed that the need for retrospective literature goes back a minimum of five years for more than 75% of our user chemists.

Another approach was to subdivide the file on a subject basis. This is supported by the fact that specialization in knowledge has advanced to such a degree that no one chemist is potentially interested in all the published material in every field of chemistry. It has been noticed that the rapid growth in science has been paralleled by a rapid increase in the number of chemists, as well as a proportional increase in the topics of chemistry. This has led the primary journals to divide into more specialized journals so that they can serve a more focused audience. Faced with a rapid increase in the coverage of the literature of specialized subjects, secondary literature sources have also decided to subdivide their contents into specialized sections, assuming that users' interests would be localized in these sections. Such localization of users' interests has been the subject of several recent studies. Lancaster, in his Medlar's evaluation study,

showed that 20% of the journals cited accounted for 75% of the retrievals
[3]. This tendency of user interests to be localized presented some
support for subdividing the file on a subject basis without affecting the
recall value of the system.

The contents of <u>Chemical Abstracts</u> are divided into 80 sections which
are grouped into five main subgroupings: Biochemistry, Organic Chemistry,
Macromolecular Chemistry, Applied Chemistry and Chemical Engineering
and Physical and Analytical Chemistry. The five section groupings provided
a convenient, predetermined way of subdividing the Condensates retrospective file. This allowed the users to have the choice of searching one or
more of these subfiles according to their subject interest. In addition, it
allowed the PCIC to structure its pricing on the basis of the subfiles instead of the whole file.

In dividing the file into separate subfiles based on the <u>CA</u> subject
groupings, the two main hypotheses developed were 1) for any one search
a comparatively few groupings (subfiles) will account for a large percentage
of retrievals, and 2) users of the file have some experience and knowledge
of the subject contents of <u>CA</u> groupings which will enable them to predict
which groupings (subfiles) will account for the greatest percent of their
output, i.e., which subfiles are most related to their questions. These
two hypotheses were investigated during the PCIC retrospective search
experiment, and will be discussed in a later part of this chapter.

A.   Construction of the File

The process of constructing the retrospective file consisted of the
following three main steps: 1) separating the documents which belong to
each subfile (Biochemistry, Organic, etc.) from each current-awareness
tape and merging like sections into a single tape; 2) converting the records
on each subfile into TEXT-PAC retrospective search format; and 3) testing of the data base and the search programs.

The separation and merging of documents belonging to each subfile is illustrated by a schematic diagram in Fig. 1. Total computer timings and
costs of the sectioning of the file for Volume 73, using an IBM 360/50 computer with 256k of core running under OS with MVT and HASP II System, is
given in Table 2. Cost is given in terms of computer usage units which
are a function of CPU time, I/O time, and memory load. One unit is approximately equal to one CPU hour and 128k of high speed core storage.

The conversion of the records on each subfile into TEXT-PAC retrospective search format is illustrated in Fig. 2. The conversion involves
a series of TEXT-PAC programs which are executed in two separate computer runs which will be referred to as Convert I and Convert II. Convert I
consists of two programs. The first (CVRT) converts <u>CA Condensates</u> data

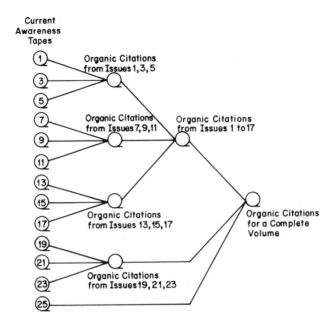

FIG. 1.   Separating and merging of citations belonging to the Organic Chemistry subfile.

into TEXT-PAC Condensed format, and the second is an edit program (TRC 260) which reformats input data into normal text format required by the system. In addition, the edit program checks the input data for key-punching and sequence errors. Convert II consists of two programs, the edit convert program TRC 210 and the retrospective merge program TRC 251. The TRC 210 program processes the edited Condensed TEXT-PAC output into two sets of records: 1) searchable records in which the words are sorted alphabetically by word length, and 2) 360 condensed text records organized by print control within document ID. The TRC 251 program merges the search and the condensed text records to produce the retro file master (the retrospective searchable tape). An example of computer timings and cost of converting the file is given for Volume 73 of CA Con-densates in Table 3. It should be mentioned that each volume of each sub-file has been recorded on a separate magnetic tape. This means that for each subfile (Biochemistry, Organic, etc.) there are five separate tapes for Volumes 69 to 73, giving a total of 25 separate tapes. Table 4 gives the number of citations on each of these tapes.

Testing of the system was done in two steps. First, some live ques-tions were run against the data base to test the search programs. Then,

TABLE 2

Computer Time and Cost of Separating and Merging
Documents Belonging to Each Subfile for Volume 73[a]

Section	Number of documents	Execution time (min)	CPU time (min)	Computer usage units[b]
Biochemistry	42, 706	71. 1	16. 9	0. 7455
Organic	16, 412	49. 3	10. 4	0. 5177
Macromolecular	14, 270	31. 3	6. 9	0. 3317
Applied and Chemical Engineering	23, 953	50. 3	11. 6	0. 5292
Physical and Analytical	39, 687	86. 5	19. 1	0. 9037
Total	137, 028	288. 5	64. 9	3. 0278

[a]These figures were derived using an IBM 360/50 computerwith 256k of
core running under OS with MVT and HASP II System.

[b]Units are the basis for computing computer cost.

Units = [(CPU time + 0. 85 IO time) • (HS core + 1/2 LCS)/128k]

some tests were run to check the contents of each subfile constructed from
the current-awareness tapes to be certain that all data were present on the
retrospective tapes. This was done by synthesizing some questions (strategies) to search for certain citations which were cited by the hard copies
of Chemical Abstracts and which were assumed to be present on the constructed retrospective tapes. The results of the test showed that there were
some gaps on some tapes as well as mixed citations (belonging to different
subfiles) on others. These errors resulted from Computer Center errors
such as merging wrong tapes or from not merging the tapes in the right
sequence, and were mainly due to the confusion caused by using a large
number of current-awareness tapes to construct the file. When this problem
emerged, a routine was written to list all abstract numbers on each merged
subfile to determine with accuracy the erroneous citations on the corresponding tapes. The listings were then checked against the range of abstract
numbers corresponding to the different section groupings in the hard
copies of Chemical Abstracts. The correction of these errors was a very
costly and time-consuming operation. It involved writing a special program

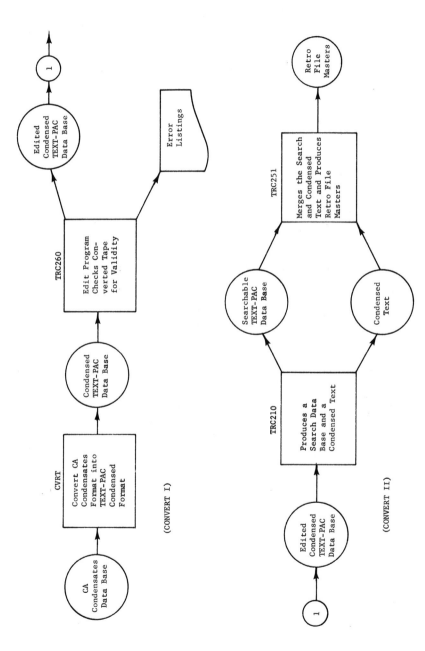

FIG. 2. Flow chart of CA Condensates to TEXT-PAC retrospective data base conversion.

TABLE 3

Computer Time and Cost of Converting Volume 73

Section	Number of citations	Convert I			Convert II			Total (Convert I & II)		
		Exec time (min)	CPU (min)	Computer usage units	Exec time (min)	CPU (min)	Computer usage units	Exec time (min)	CPU (min)	Computer usage units
Biochem-istry	42,706	172.8	145.8	5.8195	170.2	113.9	2.8785	343.3	259.7	8.6980
Organic	16,412	73.9	59.2	2.4819	69.9	46.4	1.1796	143.8	105.6	3.6615
Macro	14,270	67.8	55.4	2.2750	66.1	44.5	1.2072	133.9	99.9	3.4822
Applied and Chem. Eng	23,953	100.8	81.6	3.3868	101.1	68.4	1.8499	201.9	150.0	5.2367
Physical And Analytical	39,687	167.6	138.2	5.6357	165.3	111.8	2.7990	332.9	250.0	8.4347
Total Sections	137,028	582.9	480.2	19.5989	572.6	385.0	9.9142	1155.5	865.2	29.5131

Units = (CPU time + 0.85 IO time) • (HS core + 1/2 LSC) /128k

TABLE 4

Number of Citations of <u>CA</u> <u>Condensates</u> Retrospective Tapes

Volume \ Section	Bio.	Org.	Mac.	Appl.	Phys.	Total
69	28,171	16,659	10,401	18,706	39,048	112,985
70	34,759	17,850	11,812	21,394	35,979	121,794
71	37,930	16,951	14,430	24,087	37,775	131,173
72	41,106	17,950	14,513	25,250	40,515	139,334
73	42,706	16,412	14,270	23,953	39,687	137,028
Total	184,672	85,822	65,426	113,390	193,004	642,314
Percent	28.7	13.4	10.2	17.7	30.0	100.0

Note steady increase in the size of each volume showing increasing rate of growth of the literature.

to add or delete any range of citations on the tapes, and in many cases it was necessary to reconstruct the retrospective tapes completely (including merging and converting the current-awareness tapes from the beginning).

During our testing of the data base it was found that some citations present in the hard copies of Chemical Abstracts were not included in the current-awareness tapes received from CAS. The average number of total citations missing from the tapes of each volume (26 tapes) was 120 out of an average total per volume of 128,462 (with the exception of volume 69 which was missing 413 abstracts). Chemical Abstracts Service indicated that the reason for the discrepancy between the hard copies and the tapes was that different operations are involved in processing the tapes and the hard copies. This problem will be removed when CAS completes the implementation of its plans to mechanize the manual work flow in processing its data base.

B. File Searching

The TEXT-PAC retrospective search subsystem is designed for searches in which comparatively small numbers of searches can be conducted simultaneously against a comparatively large data base. For each retrospective search run, a new file of questions is generated. This file has a limit of roughly 200 questions depending upon their length. A flow diagram for retrospective searches is given in Fig. 3. Question cards

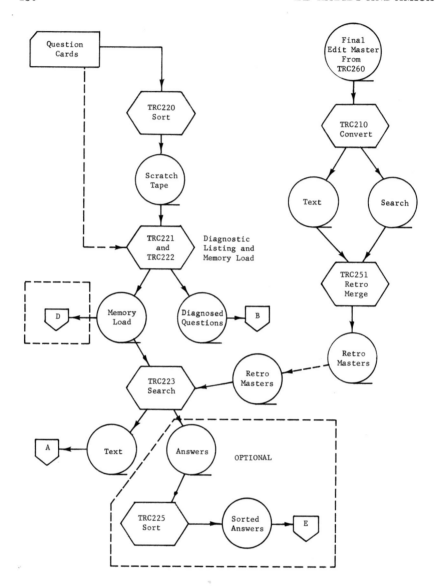

FIG. 3.   Retrospective search flow diagram.

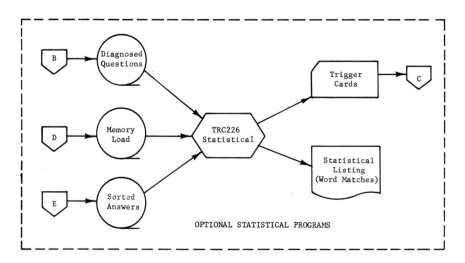

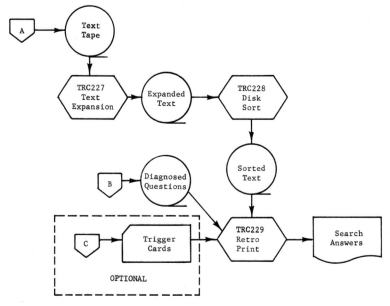

FIG. 3. (continued)

are processed and sorted by question number using TEXT-PAC program
TRC 220. The questions are entered into the system through the Retro-
Question Diagnostic program (TRC 221) which checks for format violations
and logical errors. Only correct profiles are processed into memory
loads for input into the Retro search. The Retro-Memory Load program
(TRC 222) prepares a single memory load from the questions. Once the
capacity of a memory load is reached, the search run stops and any excess
questions must be separated and resubmitted for processing in a subsequent
run. The Retro-Search Program (TRC 223) processes the Retro-Master
tape against the memory load to create a retro-answer text tape containing
the text of the answer preceded by the question numbers to which the text
relates. This is processed by a Text Expansion program (TRC 227) to
create an output file with the document text written once for every question
for which it was an answer. The Retro-Text Sort program (TRC 228) then
orders the text by question number. The sorted text and the diagnosed
questions are input to the Retrospective Print program (TRC 229), which
prints the questions and matching answers.

The retrospective search subsystem contains an optional statistical
program, primarily an aid to the user in showing what word matches and
what logic caused each answer. This helps the user in improving his questioning techniques. The program produces two types of output:
1. A Retro-Statistical Listing showing the question card images and
the number of hits. In addition, if statistics had been requested, it shows
the word matches related to each document hit and the logic involved where
possible.
2. A corresponding trigger card for each document for which a statistical listing was requested. The cards can be used to eliminate or select
printing of documents indicated to be hits by the search, but proved not to
be relevant after an analysis of the retro-statistics or documents.

C. File Update

The update of the retrospective search file is done on a current weekly
basis. A flow diagram for the file update is given in Fig. 4. Current-
awareness tapes received from CAS are sectioned so that the records which
belong to each subfile are separated and recorded on separate magnetic
tapes. The data on each tape is then converted into the TEXT-PAC retrospective search format. Both Convert I and Convert II are executed in
one single computer run using the TEXT-PAC conversion programs mentioned previously. The converted data on each tape is then appended by a
general utility routine (IEBGENER) to the corresponding old retrospective
master tape. Although the update is done weekly, the actual update for
each subfile is completed every two weeks, since the complete contents of
CA Condensates are published in two alternative weekly issues. The weekly
current update of the file permits the user to search the file up till the most
recent issue published by CAS.

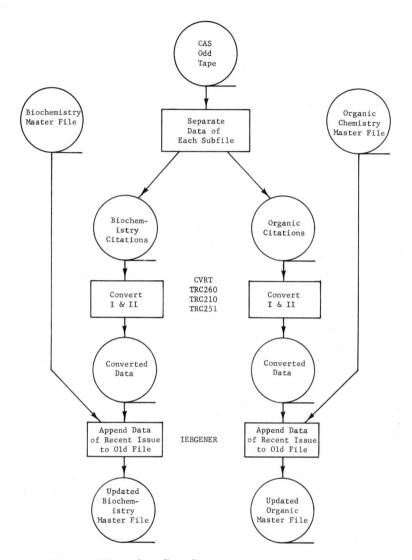

FIG. 4.   File update flow diagram.

III. THE EXPERIMENT

A. Objectives

The basic goal of the retrospective search experiment was to test the feasibility and application of large-scale retrospective searches on CA Condensates. Our aim was to determine by the end of the experimental period either a) that mechanized retrospective searching of Condensates is practical, economical, and desirable for various types of users, or b) that machine searching of large files is not feasible for any or all of these reasons, and that further research on this type of service is impractical with presently available systems.

During our experimentation with the current-awareness service, our users frequently indicated a strong interest in computer-based retrospective searching. An important objective of the retrospective search experiment was then to determine the extent of such interest among chemists, and study their needs for using the service. The main purpose of studying the user's needs was to determine the basic characteristics of information requirements for the retrospective user groups. Our criteria to define these needs was to find the answers to the basic questions faced by the system designer, which are:

1. Who wants retrospective searching and what do they want it for? Does the chemist use the service as a tool to acquaint himself with a new research area, or to refresh himself about areas with which he is already familiar? Are there other reasons?
2. What form of output do chemists desire?
3. What is the direct functional usefulness of retrospective searches? At which stage of a research project would a retrospective search be most helpful to the chemist?
4. How far back in the literature do users want to go?
5. How much are users willing to pay for the service, given their present expectations and sizes of budgets within which they must work?
6. How long will users be willing to wait for results?

We believe that the answers to these questions are not only important for the success of our system, but also for the future application of computer-based chemical information and its development. The articulation of the discovered information needs and uses with the planning and design activities is fundamental to the success of any mechanized system. They are necessary for the manipulation of the essential conditions needed to control and improve the utilization of information.

It was equally important to the success of the system to determine and characterize the expectations which scientists have of the retrospective search system. After identifying the retrospective search needs of the users, our objective was to evaluate the degree to which the retrospective search system had satisfied these needs. Related to this evaluation were such questions as the user's impression of the service, whether he recommended it to others, and whether he is willing to use it again. These questions will be dealt with in detail later in the chapter.

One of the most important issues in developing the system was the question of economics which has a direct relationship to two main elements: the size of the file, and the search system. We indicated before that our approach to the problem of the size of the file was to subdivide the data base on a subject basis. One of the objectives of our experiment was to test the hypotheses that led us to divide the file; a) that few subdivisions will account for a large percentage of the retrievals for a given profile in a given area, and b) that users of the file will be able to predict which subdivisions will account for the greatest percent of their search output (see Section II).

Another objective of the experiment was to determine the economic performance of TEXT-PAC as a retrospective search system. Little work has been done so far on the evaluation of retrospective search systems from an economic point of view. Our objective was to collect data on a) the processing cost of constructing the file and b) the search cost of the file and its relationship to the batch size of different queries, number of terms associated with them, size of the subfile searched, and number of alerts retrieved.

A major factor governing the performance of a mechanized search is the quality of interaction between the user and the system. The process of transforming an information need into a formal system request becomes more critical for free-text material. It involves an element of speculation as to the manner in which concepts are expressed in the file. This becomes quite serious in retrospective searching, especially of such large files as the Condensates.

Another objective of the experiment was to explore the best means to search the file and to see whether a preliminary search conducted against a small portion of the file would help the user in preparing a better profile.

In summary, the principal objectives of the retrospective search experiment were:

1. To study the search requirements and needs of machine retrospective searching of chemists.
2. To evaluate how effectively our system fulfills these needs.
3. To evaluate the feasibility of subdividing the file on a subject basis as an economical approach for machine retrospective searching.

4. To determine the economic performance of the system and to
    evaluate TEXT-PAC as a retrospective search system.
5. To determine the best means for users to explore the data base.

B. Experimental Design

In the designing of an appropriate experimental system that would serve
our experimental objectives, the most critical problems faced were 1) to
select a suitable experimental user group, 2) to determine the procedure
for obtaining the required data on user needs, and 3) to establish the method
for evaluating the subdivision of the file on a subject basis.

1. Selection of the Experimental User Groups

A large part of the effort that went into the test design was devoted to
selecting the user groups. Initially it was thought that the test group for
retrospective searching should be a subset of the total set of the PCIC users
who participated in the current-awareness experiment. With the large
amount of demographic and professional information collected from the
PCIC users, it should have been possible to identify potential retrospective
search users from the general user population without further data collection. Further, it would have been possible to compare the results from
retrospective searches with bibliographies already developed by users,
including records from their current-awareness service. However, when
the current-awareness users were invited to submit profiles for retrospective searching, their initial response was very poor. It appeared that this
lack of interest was due to one of the following reasons: a) that most of our
users had already satisfied their current information needs from the current-awareness searches and decided that retrospective searching would
provide them with redundant information; or b) that the interest expressed
by some of our users in machine retrospective searching was artificial,
and that chemists have little real desire for such a service. It was thought,
however, that the first reason was more logical than the second since the
majority of the users had joined the current-awareness service from its
beginning, and since the retrospective file consists of the same tapes
searched for the current-awareness information.

To confirm our hypothesis, it was decided to extend the participation
in the retrospective search experiment to a fresh sample of users from the
same organizations that participated in the current-awareness service as
well as from other organizations. In addition, an attempt was made to
obtain the participation of a second user group from Great Britain. The
addition of the latter user group allowed us to observe whether user needs
of chemists in the United States are significantly different from those of

Great Britain, and to test the results and conclusions derived from the primary sample in the United States. The participation of the users from Great Britain was arranged through the United Kingdom Chemical Information Service (UKCIS) at the University of Nottingham. All the direct liaison with the British users was carried out by the Information Specialist at UKCIS.

Test requests were taken completely at random as they were presented to the system by users who were invited to participate in the experiment. The experimental program was made known to users in the United States through workshops, a direct mail program and personal visits from marketing representatives of the Knowledge Availability Systems Center. The primary method of inviting users in Great Britain to participate in the experiment was by telephone calls from the UKCIS information staff.

Users from both samples were composed of scientists representing the major sectors of the chemical community, i.e., a stratified sample of scientists from universities, nonprofit research institutes, industry, and government.

The sample of the American users who responded to the invitation to participate in the program is composed as follows:

	No. of users	Percent	No. of profiles	Percent
Universities	73	40.1	92	42.0
Industry	87	47.8	98	44.7
Nonprofit Research	18	9.9	25	11.4
Government	4	2.2	4	1.9
Total	182	100.0	219	100.0

The sample from Great Britain is composed as follows:

	No. of users	Percent
Universities	28	31.1
Industry	40	44.5
Government	22	24.4
Total	90	100.0

We can notice that the group from Great Britain has proportionally more government scientists and fewer university researchers than the U.S. group.

The sample from Great Britain was much more controlled than the
U.S. sample. It was selected from users who had participated in other
experimental studies conducted by UKCIS and who were known to be able to
provide good profiles and intelligent feedback.

2. Method of Investigating User Needs

In order to collect the behavioral data on user needs and expectations,
two primary methods of data collection were used. Each user who submitted
a retrospective search request was sent a questionnaire and asked to complete it and return it by mail before his profile was searched. The purpose
of this questionnaire was to probe the user's expectations of the service.
Details of the questionnaire are given in the section on Experimental Results.

The second method of behavioral data gathering was a telephone interview designed as a follow-up to the questionnaire. The interview required
about five minutes and was held after the user had received his retrospective search output and had had an opportunity to evaluate and act upon it.
The purpose of the interview was to determine the degree to which the
retrospective search had lived up to the users' expectations, how the users'
restrospective information needs related to their current-awareness information needs, the relevance of the output to their information needs, and
how the output would effect various stages of their research. It consisted
of the following questions:

1. How did you become aware of the availability of the retrospective
   literature searching service?
2. What is the general impression of the output that you received?
3. Was the subject area of your retrospective search related to the
   subject area of your current-awareness search?
4. At what stage of the research (early conceptualization, experimentation, writing) covered by your retrospective search, are you
   presently engaged?
5. For which stage of research will your retrospective output have
   its greatest functional usefulness?
6. Of the total output which you received, please estimate the percentage of the alerts that were relevant to your interests.
7. When you find a relevant alert, how do you generally follow it up?
   (Chemical Abstracts, reading the original article, reprint)
8. Do you rely very heavily on Chemical Abstracts as a literature
   tool? In what ways?
9. Have you had to supplement your computerized retrospective literature search with a manual search of the literature?
10. For the particular research program covered by your retrospective
    search, how far back in the literature will you have to search?

11.   Do you think that you will have need of another retrospective search in the future ?

12.   Have you recommended the use of our retrospective search to any of your colleagues ?

13.   Can you estimate the value of this retrospective literature search to you in terms of money, i.e., how much you would be willing to pay for what you received ?

The interview used for the British group was slightly modified to account for their different circumstances.

3.   Method of Evaluating the File Subdivision

The subdivision of the file as an economic approach to retrospective searching was based on the assumption that the material on the topic of interest, at least insofar as it is definable, is as localized as possible in those sections of the file to which the inquiry is addressed, As we mentioned previously, we hypothesized that a) for each search inquiry the localization of pertinent references occurs in comparatively few subdivisions of the Condensates file, and b) the user of the file, whether he is the requester or the information analyst, is able to determine in advance the subdivisions in which the material is localized.

The evaluation of the file subdivision was based on testing the above-mentioned hypotheses in relation to the search requests made by the test group in a 12-month period.   The testing of the hypotheses was done by measuring the ratio of the results a user can obtain by searching one, two, three, four, or five subdivisions of the file.   These ratios were then compared with those obtained from searching the file subdivisions which the user predicted to contain most of the information relevant to his search request.   The procedure followed in conducting the test is given below.

The users were informed in advance that the file had been broken into the five main subdivisions of Chemical Abstracts and that each of these subdivisions could be searched separately depending on their interests.   Each user was asked to complete a profile entry form on which he indicated the subdivisions he wished to search and are pertinent to his search request. In addition, each user was asked to give a complete statement of the problem or the question for which his search was to be developed and also a list of suggested terms (key words, authors, names of compounds) which described his interests and could be used to form the search strategy.   The search strategy was then prepared by the information analyst after negotiating the request with the user.   In determining the file subdivisions pertinent to his search request, the user had the opportunity of seeking the advice of the information specialist at his organization or at the PCIC, especially if he had little experience with Chemical Abstracts.   It must be stressed here that such assistance given by the information specialist to the user does not defeat the purpose of the experiment (testing the hypothesis that pertinent subdivisions can be determined in advance) since in a real operating situation the information specialist acts as the intermediary between the user and the system, and is available for such consultation.

For each of the test requests received at PCIC, a search was conducted against all the five subdivisions of the Condensates file. Since there were five separate tapes for Volumes 69 to 73 for each file subdivision, it took 25 computer runs to complete the searches for each profile. Computer output from each separate search was delivered to the users as it was completed without awaiting the remaining searches. When the user received his first search output he was informed of the purpose of the experiment, and that in spite of his interest in only a few subdivisions his request would be searched against all subdivisions so that results could be compared and evaluated. Users were informed from the beginning, before their requests were accepted, that the service was experimental and that feedback information, necessary for evaluating the results, would be sought from them. Accordingly, each user was asked to determine the number of relevant citations (hits) in the output he received from each separate search and how many hits were known to him prior to receiving the search results. A copy of the form used by the user to evaluate the results from each search is shown in Fig. 5.

In conducting the test for the sample from Great Britain a slightly different procedure was followed in entering the user profiles and searching them against the file. While the retrospective search service was offered completely free of charge for the U.S. users, their British counterparts were charged £10 ($26) per profile by UKCIS to cover the expenses of coordinating the test with PCIC. Each user from Great Britain was asked to provide his profile written in UKCIS search profile standard format with which they were all familiar. In addition, a covering description in natural language was required. The profiles were then converted into the TEXT-PAC profile standards.

The sample of profiles from Great Britain was used to evaluate the extent of improvement of the interaction level between the user and the file by means of conducting a preliminary search against a small portion of the file. Each of these profiles was searched against Volume 73 (the latest volume) of one of the file subdivisions for which the search was requested by the user. Results of the preliminary search were then delivered to the user for him to evaluate and decide on any changes or amendments for his profile before it was searched against the rest of the file.

4. Derivation of a Statistical Method for Evaluating the File Subdivision

Search results for each profile and the results of user evaluations were tabulated as shown in Table 5. As can be seen from the table, each

University of Pittsburgh
Pittsburgh Chemical Information Center

CA Condensates Retro Search Results

Profile No.   530016

Company Code

Enclosed are the results of a chemical condensates retrospective search of

      Volume No. ____73_____

      Section _____B10_____

For which ____73_____ references are cited. Please indicate below

1) The number of citations which you consider to be of value to the
   problem that prompted your search request _____11_____.

2) How many of those references you indicated in (1) above, which you
   considered of value to your problem, were known to you prior to
   receiving these search results _____4_____.

Please use the space below for any comments, recommendations, or criticisms
you may have of this service.

Please return this sheet to:

      Bahaa El-Hadidy
      Pittsburgh Chemical Information Center
      Mellon Institute, Room 503
      Pittsburgh, Pa. 15213

"I have found this service to be most helpful in my initial literature

search. It has been comprehensive yet not overloaded with extra

material. I was most interested to find a good percentage of articles

that I had either overlooked or dismissed at first glance and which

were closely related to my field of interest.

Please continue this service--if there is any cost in the future for

new searches, I would be willing to pay - within limits."

FIG. 5. User's feedback form.

TABLE 5

Recording of Search Results and User Feedback Results for Each Profile

DATE STRATEGY RECEIVED ___3-10-71___

VOLUME	① BIO HITS	DATE	③ ORG HITS	DATE	⑤ MAC HITS	DATE	② APP HITS	DATE	④ PHY HITS	DATE
69	(a) 152 / (b) 6 / (c) 0	4-12-71	5 / 0	4-9-71	0 / 0	3-22-71	3 / 0	3-19-71	1 / 0	4-12-71
70	202 / 15	3-29-71	0 / 0	4-9-71	0 / 0	3-17-71	2 / 0	4-3-71	6 / 0	3-25-71
71	200 / 12	4-12-71	3 / 0	4-13-71	0 / 0	3-29-71	3 / 0	3-18-71	2 / 0	3-15-71
72	200 / 80	4-15-71	3 / 0	4-13-71	0 / 0	3-29-71	4 / 0	3-15-71	0 / 0	3-10-71
73	225 / 30	4-6-71	2 / 0	4-16-71	0 / 0	4-12-71	6 / 4	4-16-71	1 / 0	3-10-71
TOTAL	979 / 143 / 8		13 / 0 / 0		0 / 0 / 0		18 / 4 / 0		10 / 0 / 0	

DATE SERVICE COMPLETED ___4-16-71___
(a) Number of citations retrieved (alerts)
(b) Number of relevant citations (hits)

NUMBER OF TERMS ___32___

PROFILE NUMBER ___520016___

column contains the results pertaining to each file subdivision (subfile) which include: a) the number of citations (alerts) retrieved by searching the profile against each volume, b) the number of relevant citations (hits) which the user considered to be of value to the problem that prompted his search request, and c) the number of hits known to the user prior to receiving his search request.

When all the searches against all tapes were completed for the profile and all the evaluations were received from the user, the results from all five volumes (69 to 73) pertaining to each subfile were totaled. The subfiles were then ranked in decreasing order with regard to the total number of alerts retrieved by each subfile. For example, in Table 5, for profile No. 520016, the five subfiles were ranked so that Biochemistry, which retrieved the highest number of alerts from all volumes, was ranked first; Applied, second; Organic, third; Physical fourth; and Macromolecular, with the lowest number of alerts, was fifth.

Having ranked the subfiles according to the localization of references pertinent to the search inquiry, the number of alerts and hits for each subfile were tabulated according to their rank as shown in Table 6, which gives the search statistics for profile No. 520016. This allowed the derivation of the following figures.

1. The percent of alerts retrieved by each of the ranked subfiles P(L) which is computed as follows:

$$P(L)_k = \frac{N_k}{N_{tot}} \times 100$$

where N_k = number of alerts retrieved by subfile k

N_{tot} = total number of alerts retrieved by all subfiles

$$(N_1 + N_2 + N_3 + N_4 + N_5)$$

For example in Table 6 the percent of alerts P(L) for the Applied subfile would be:

$$P(L)_{applied} = \frac{18 \times 100}{979 + 18 + 13 + 10 + 0} = \frac{1800}{102} = 1.7.$$

2. The accumulative percent of alerts retrieved through each of the ranked subfiles AP(L) which is computed as follows:

$$AP(L)_k = \frac{100 \sum_{i=1}^{k} N_i}{\sum_{i=1}^{5} N_i} = \frac{100 \sum_{i=1}^{k} N_i}{N_{tot}} .$$

TABLE 6

Retro Searches Statistics for Profile No. 520016

Section	1 Biochemistry	2 Applied and Chem. Engineering	3 Organic	4 Physical and Analytical	5 Macro-molecular	Total
Total citations searched	184,672	113,390	85,822	193,004	65,426	642,314
Percent	28.7	17.7	13.4	30.0	10.2	100
Alerts						
Number of alerts	979	18	13	10	0	1020
Percent	96.0	1.7	1.3	1.0	0	100
Accumulative percent	96.0	97.7	99.0	100	100	100
Hits						
Number of hits	143	4	0	0	0	147
Percent	97.3	2.7	0	0	0	100
Accumulative percent	97.3	100	100	100	100	100
Precision	14.6	22.2	0	0	0	14.4
Accumulative precision	14.6	14.7	14.6	14.4	14.4	14.4

The accumulative percent of alerts for the Biochemistry, Applied, and Organic subfiles in Table 6 would then be:

$$AP(L)_{organic} = \frac{(100) \times (979 + 18 + 13)}{1020} = \frac{100 \times 1010}{1020} = 99.0.$$

3. The percent of hits (relevant alerts) retrieved by each of the ranked subfiles P(H) which is computed as follows:

$$P(H)_k = \frac{M_k}{M_{tot}} \times 100,$$

where M_k = number of hits retrieved by subfile k and M_{tot} = Total number of hits retrieved by all subfiles $(M_1 + M_2 + M_3 + M_4 + M_5)$.

The percent of hits for the applied subfile in Table 6 would be:

$$P(H) = \frac{4 \times 100}{143 + 4 + 0 + 0 + 0} = \frac{400}{147} = 2.7.$$

4. Accumulative percent of hits retrieved through each of the ranked subfiles AP(H) which is computed as follows:

$$AP(H)_k = \frac{100 \sum\limits_{i=1}^{k} M_1}{5 \quad M_i \atop i=1} = \frac{100 \sum\limits_{i=1}^{k} M1}{M_{tot}}.$$

The accumulative percent of hits through the Biochemistry, Applied, and Organic files in Table 6 would then be:

$$AP(H)_{organic} = \frac{(100) \times (143 + 4 + 0)}{147} = \frac{14700}{147} = 100.$$

5. The precision of each of the ranked subfile (R) which is the percent of the total articles judged of value to the total articles assessed in that subfile. This is computed as follows:

$$R_k = \frac{M_k}{N_k} \times 100.$$

For example, the precision of the Biochemistry subfile in Table 6 is:

$$R_{bio} = \frac{143}{979} \times 100 = 14.6.$$

6. The accumulative precision through each of the ranked subfiles (AR) which is computed as follows:

$$AR_k = \frac{\sum\limits_{i=1}^{k} M_i}{\sum\limits_{i=1}^{k} N_i} \times 100.$$

The accumulative precision through Biochemistry, Applied, and Organic subfiles in Table 6 would then be:

$$AR_{org} = \frac{143 + 4 + 0}{979 + 18 + 13} \times 100 = 14.7.$$

Having calculated and recorded the values of the percent and the accumulative percent of alerts and hits as well as the precision and accumutive precision for all the users who completed their evaluations for the search output, the next step involved the derivation of the means and standard deviations for each of these values for the whole sample.

On the basis of all these records it was possible to compute the average percent of alerts and hits retrieved by one, two, three, four, or all the five subfiles. This was used as our measure to asses the results a user can get by searching one or more subfiles in the file and evaluate the extent of localization of references in the file subdivisions.

5. User's Ability to Predict Relevant Subfiles

The ability of users to predict relevant subfiles was evaluated by two methods. First, the user's selections (predictions) of pertinent subfiles were matched against the actual pertinent subfiles which were ranked according to the localization of references which are pertinent to the search inquiry. (As mentioned previously, each user was asked to select in advance the subfiles he wished to search and are pertinent to his search request.) For each ranked subfile, a match occurred when the user selection matched the ranked subfile. For example, if a user requested to search two subfiles, say the Biochemistry and Organic ones, and these same two subfiles retrieved the highest number and the second highest number of alerts, then a match was recorded for both the first and second ranked subfiles. Had the results indicated that only one of the user's selections, say the Organic subfile, retrieved the highest number of alerts, while another, unselected subfile, such as the Physical chemistry one, retrieved the second highest number of alerts, then only a match would have been recorded for the first ranked subfile, and there would have been no match for the second ranked subfile. Second, a comparison was made for the percentage of alerts and hits retrieved by the users requests to search specific subfiles and the

actual percent of total alerts and hits (for the ranked subfiles) which are
retrieved by the whole data base.

IV. EXPERIMENTAL RESULTS

A. User Needs

The behavioral experiment with the users of the retrospective literature
searching service was designed to determine and characterize the expectations which scientists have of such a system. We wanted to identify the
retrospective search needs of the users and then proceed to evaluate the
degree to which the retrospective service satisfied those needs.

As mentioned previously, there were two samples of users involved, a
primary user group in the United States, and another user group located in
Great Britain. In order to determine user expectations of the service, each
person who submitted a retrospective search request was sent a questionnaire and was asked to complete it and return it by mail. As of December
15, 1971, 164 U.S. users and 114 British users had returned their questionnaires providing respectively a 90% and a 92% response rate for both
groups. (In addition to the 90 users who submitted profiles, some information specialists in the organizations whose scientists participated
in the experiement responded separately. These were treated as users,
thus in the final analysis we have 114 respondent users.)

The results of the questionnnaire follow, and some interesting comparisons can be made between the scientists in the United States and Great
Britain. The reader is reminded of a somewhat different composition of
the two samples as presented earlier.

1. Why have you requested a retrospective search?

U.S.		G.B.		
n	%	n	%	
53	33	30	27	a. To become acquainted with a new re-research area
33	20	15	14	b. To refresh myself regarding a research area with which I am only slighty familiar
77	47	60	54	c. To complete the coverage of a research area with which I am already very familiar
0	0	6	5	d. Other (please specify)
163	100	111	100	

Notice that a slightly higher percentage of the U.S. users are using the service as a tool to acquaint themselves with new research areas. The six G.B. users who chose the "other" category expected to use the service solely for the purpose of evaluating its effectiveness as a literature tool. In general, the two groups responded in a similar manner.

2. What form of output do you prefer?

U.S.		G.B.			
n	%	n	%		
109	66	85	75	a.	Cards
48	30	29	25	b.	Computer paper
7	4	0	0	c.	Other (please specify)
164	100	114	100		

Both groups were in agreement on their preference for cards. Those in the U.S. group choosing "other" either wanted 4 x 6 cards or 8 x 11 paper.

3. For which stage of research do you expect your retrospective search will have the greatest functional usefulness?

U.S.		G.B.			
n	%	n	%		
31	19	8	7	a.	Early conceptualization
20	12	11	10	b.	Preliminary research design
12	7	12	11	c.	Final research design
36	22	32	28	d.	Experimentation
64	40	40	35	e.	Writing reports or papers
0	0	11	9	f.	Other (please specify)
163	100	114	100		

The U.S. group had a much higher percentage of scientists who felt that the restrospective search would be most useful at the early conceptual stage of research. The G.B. scientists also has eleven users who chose the "other" category. These individuals were either not certain as to how they would use their output or they felt that it would be useful at all stages of the research process.

4. Given your current needs, how far back in the literature will you have to search?

	U.S.			G.B.			
n	%	c%	n	%	c%		
2	1	1	2	2	1	a. 1 year	
9	5	6	1	1	2	b. 2 years	
20	12	18	8	7	9	c. 3 years	
8	4	22	8	7	16	d. 4 years	
52	32	54	22	19	35	e. 5 years	
4	2	56	6	5	40	f. 6 years	
12	7	63	10	9	49	g. 7 years	
2	1	64	2	2	51	h. 8 years	
55	36	100	55	49	100	i. 9 years or more	
164	100		114	100			

It seems as though many more of the scientists in the G.B. sample have need of more dated literature (nine or more years) than do the U.S. scientists. Fifty-four percent of the U.S. scientists need to search back over a period of a least five years while only 35% of the British scientists perceive this need.

It is important to remember that the system's coverage went back through 1968, and at the time of this questioning, the users were only able to search the most recent two and one-half years of the chemical literature by computer. It should be noted that even at the three-year coverage level, the system is only satisfying 18% of the U.S. group's needs and 9% of the British group's needs. This is another example of the value of an information service having a long-latency period before its total real value is manifest. Its value will obviously increase

over the years as more and more of the chemical literature coverage is
extended. Of course, scientists beginning research in a new area will
find the present system very valuable for bringing them up to date on more
recent developments.

5. How long is a reasonable waiting period from the time you submit
    your search profile until the time you receive your output?

U. S.			G. B.				
n	%	c%	n	%	c%		
0	0	0	0	0	0	a.	1 day
20	12	12	0	0	0	b.	1 week
40	24	36	25	22	22	c.	2 weeks
43	26	62	14	12	34	d.	3 weeks
47	29	91	44	39	73	e.	4 weeks
0	0	91	6	5	78	f.	5 weeks
6	4	95	18	6	94	g.	6 weeks
0	0	95	0	0	94	h.	7 weeks
8	5	100	7	6	100	i.	8 weeks
164	100		114	100			

The sample from Great Britain seems to be more patient than the
U.S. group. Ninety-one percent of the U. S. group wanted their output
within one month while only 73% of the British group wanted their
output within this period. Again, these are not significant differences.

We can conclude by saying that, for the most part, the U.S. and
British groups were similar in the way they responded to these questions.

B. File Subdivision

A total of 240 test requests out of 309 received by PCIC were searched
against all the five subfiles of CA Condensates. Of these, 150 (out of 219)
pertain to the U. S. group, and 90 pertain to the British group. The remaining 69 requests of the U. S. users, received between June and September 1971, were searched only against the subfiles the users indicated to be
pertinent to their search requests.

Operational difficulties caused us to limit the searches for 69 requests
of the U. S. group to the subfiles requested by the users. A large

percentage of the search requests made by the U. S. group was received
a relatively short time after it was decided to extend the participation in
the experiment to a large number of organizations. Due to the enormous
amount of computer time required for searching all subfiles, and since
the computer used in searching the file was not dedicated to such an operation, a large backlog of searches was created by June 1971. At the same
time the search requests from the British group were received. Since the
number of requests received from the U. S. group by that time was sufficient for testing the file subdivision, we decided that all requests received
later than June 1st would be searched only against the pertinent subfiles.
However, by September the situation had improved and searches on all
subfiles were resumed for the U. S. users.

As of this writing, a total of 113 sets of evaluations of searches for
U. S. users and 61 sets of evaluations for British users were returned,
providing a 75.3% and 67.8% response rate, respectively, for each
group. The comparatively lower percent response rate for the
British group is probably due to the fact that their searches had only been
completed shortly before this writing and results were still being evaluated
by the users. The distribution of the sample according to type of organization is given in Table 7.

TABLE 7

The Sample

(1) U.S.	No. of organizations	No. of profiles		Responded	
		N	%	N	%
Universities	10	51	34.0	36	31.9
Nonprofit research	2	10	6.7	6	5.3
Government	2	2	1.3	1	0.9
Industry	25	87	58.0	70	61.9
Total	39	150	100.0	113	100.0
(2) British		N	%	N	%
Universities	11	28	31.1	12	19.7
Government	10	22	24.4	18	29.5
Industry	24	40	44.5	31	50.8
Total	45	90	100.0	61	100.0

Response Rate: U.S. 75.3%   British 67.8%

The results for 17 U.S. and 13 British profiles were ignored for the following reasons:

1) Twenty-two profiles had no or few alerts, i.e., the computer searches for these profiles showed that none or few of the documents in the file were pertinent to the search strategies of the profiles;

2) eight profiles had no or few hits, i.e., the user evaluations for the search outputs of these profiles showed that none or few of the documents retrieved were considered of value to their problems.

The results of these 30 profiles were not included in the final analysis conducted on the remaining 144 profiles.

1. Percent of Alerts and Hits Retrieved by the Subfiles

When the individual accumulative percents of alerts retrieved through each of the ranked subfiles AP(L) for the U.S. profiles were averaged, we arrived at the results displayed in Table 8.

The results in Table 9 are obtained from a breakdown of the profiles according to the subject area of the subfiles that retrieved the highest number of alerts. The accumulative percents of alerts for each resulting group of profiles (which will be called a subject group) were then averaged.

Graphical representation of data in Tables 8 and 9 is given in Fig. 6 which displays the relationship between the average number of alerts retrieved and the number of subfiles searched.

The average accumulative percents of hits (relevant alerts) retrieved through each of the ranked subfiles AP(L) for the U.S. sample are presented in Table 10.

The average accumulative percents for each of the groups of profiles resulting from a breakdown of the whole sample according to the subject area of the subfile which retrieved the highest number of alerts (subject groups) are presented in Table 11.

Graphical representation of data in Tables 10 and 11 (averages of accumulative percents of hits) is given in Fig. 7.

Results of accumulative percents of alerts and hits (Tables 8 and 10) show that, on the average, a user of CA Condensates file retrieves 75.3% of references pertinent to his search request (alerts) and 78.7% of references considered of value to his search problem (hits) by searching only the single subfile of the data base which is most pertinent to his inquiry.* These percentages increase to:

*Pertinency of a subfile has been determined on the basis of the largest number of alerts retrieved in response to the inquiry. A study is being conducted to determine if the user of the file can predict in advance the subfile which would provide him with the most pertinent (and relevant) references to his search request. Results of the study will be published later.

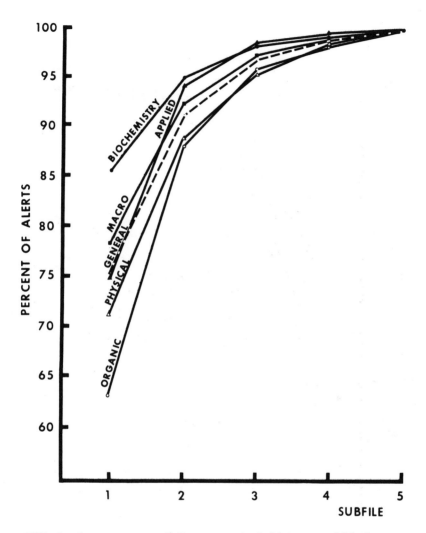

FIG. 6. Average accumulative percent of alerts per subfile for
U. S. sample.

91. 4% of alerts and 94. 5% of hits by searching the two most
pertinent subfiles,
97. 0% of alerts and 98. 5% of hits by searching the three most pertinent
subfiles,
99. 1% of alerts and 99. 6% of hits by searching the four pertinent
subfiles, and 100% of alerts and hits by searching all subfiles.

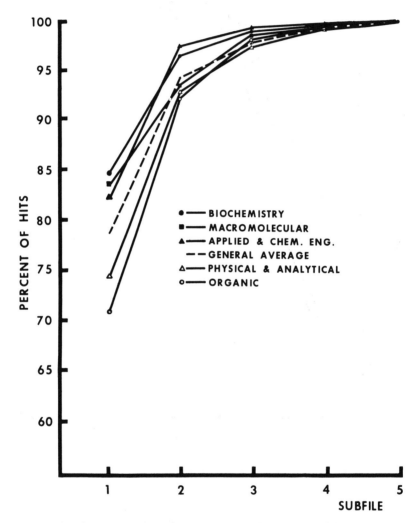

FIG. 7. Average accumulative percent of hits per subfile for
U.S. sample.

TABLE 8

Average Accumulative Percent of Alerts per
Subfile for U.S. Sample

Subfile	1	2	3	4	5
Accumulative percent (Total: 96 Profiles)	75.3	91.4	97.0	99.1	100

TABLE 9

Average Accumulative Percent of Alerts per
Subfile for Subject Groups

Biochemistry (19 Profiles)	85.8	95.2	98.6	99.6	100
Organic (12 Profiles)	63.7	88.3	96.0	98.5	100
Macromolecular (19 Profiles)	78.7	92.5	97.5	99.2	100
Applied & Chemical Engineering (12 Profiles)	75.2	94.4	98.8	99.8	100
Physical & Analytical (34 Profiles)	71.6	88.9	95.5	98.8	100

TABLE 10

Average Accumulative Percent of Hits per
Subfile for U.S. Sample

Subfile	1	2	3	4	5
Accumulative Percent (Total: 96 Profiles)	78.7	94.5	98.5	99.6	100

TABLE 11

Average Accumulative Percent of Hits per
Subfile for Subject Groups

Biochemistry (19 Profiles)	84. 8	96. 8	99. 1	99. 9	100
Organic (12 Profiles)	70. 7	92. 5	98. 8	99. 8	100
Macromolecular (19 Profiles)	83. 5	93. 7	98. 9	99. 7	100
Applied & Chemical Engineering (12 Profiles)	82. 0	97. 8	99. 5	99. 9	100
Physical & Analytical (34 Profiles)	74. 3	93. 2	97. 5	99. 3	100

In other words, while the percent of alerts increases by 16.1, 5.6, 2.1 and 0.9 as a result of searching the second, third, fourth, and fifth pertinent subfile, respectively, the percent of hits increases by 15.8, 4.0, 1.1, and 0.4, respectively. This indicates that the material on topics of interest are mainly localized in two out of the five subfiles of the data base.

By comparing the results obtained from the breakdown of the sample according to the subject area of the subfile which retrieved the highest number of alerts (Tables 9 and 11), some interesting observations can be made on the resulting subject groups.

1. The average accumulative percents of alerts and hits for the parallel subfiles vary from one subject group to another. The variations however, are much greater in the results of the first and second ranked subfiles than those of the third and fourth. This indicates that pertinent references in the file are more or less localized within a few subfiles, and that the extent of localization of material varies with different subject areas.

2. There is a marked similarity in the extent of variations between subject groups when comparing the results of average accumulative per-cents of hits and alerts. This can be observed clearly by comparing the curves of similar subject groups in Figs. 6 and 7.

3. The extent of localization of references in the file is markedly high for the Biochemistry group, i.e. for search inquiries mainly related to biochemistry.

4. The extent of localization of references in the file is comparatively low for both the Organic and Physical Chemistry groups.

Tables 12 and 13 give the results of the comparison of the average accumulative percents of alerts and hits for the U.S. and British samples. The data, represented graphically by Fig. 8, show a marked similarity in the results of the two samples.

The same similarity was noticed in the comparison of the results of the subject groups of the U.S. and the British samples, (Figs. 9, 10, 11 and 12). The only deviation was found in the results of the Applied Chemistry groups (Fig. 13). This is probably due to the small number of Applied Chemistry profiles in the British sample (5 profiles) compared to those of the U.S. sample (12 profiles).

2. Precision of Subfiles

Results of averaging the precision of each of the ranked subfiles (R) for the U.S. sample and for each group of profiles resulting from the

TABLE 12

Comparison of Accumulative Percent of Alerts
per Subfile for U. S. and British Sample

Subfile	1	2	3	4	5
U.S. Sample (96 Profiles)	75.3	91.4	97.0	99.1	100
British Sample (48 Profiles)	73.9	90.5	96.6	99.1	100

TABLE 13

Comparison of Accumulative Percent of Hits per
Subfile for U. S. and British Sample

Subfile	1	2	3	4	5
U.S. Sample (96 Profiles)	78.7	94.5	98.5	99.6	100
British Sample (48 Profiles)	79.1	94.1	98.2	99.6	100

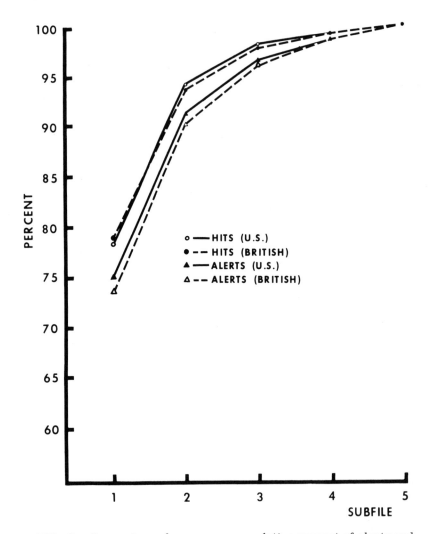

FIG. 8. Comparison of average accumulative percent of alerts and hits per subfile for U.S. and British samples.

breakdown of the sample according to the subject area of the subfile with the highest percent of alerts (subject groups) are presented in Tables 14 and 15, respectively.

Results of the average accumulative precisions through each of the ranked subfiles (AR) for the U.S. sample and for different subject groups are presented in Tables 16 and 17, respectively.

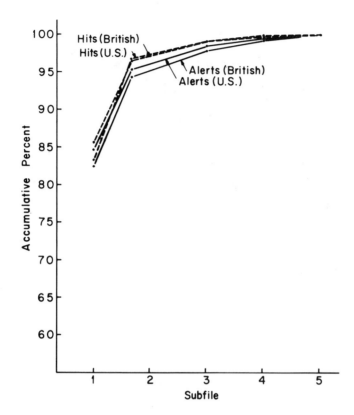

FIG. 9. Comparison of accumulative percent of alerts and hits per subfiles for Biochemical profiles of U.S. and British users.

Graphical representation of data in Tables 14 and 15 is given in Fig. 14. Figure 15 illustrates the data of Tables 16 and 17.

Table 15 shows that the average precision of the ranked subfiles decreases steadily from one to another, with the greatest variance occurring between the second- and third-ranked subfiles. The difference between the average precision of the first-ranked subfile (39.9) and the last (4.5) is notable. This indicates that out of the total references retrieved by each ranked subfile, there was a much higher percentage of relevant references in the first two subfiles than in the last three. These results validate our previous conclusion that there is high localization of pertinent references in the first two most pertinent subfiles.

Table 16 shows that the average accumulative precision through each of the ranked subfiles decreases very slightly from one subfile to another.

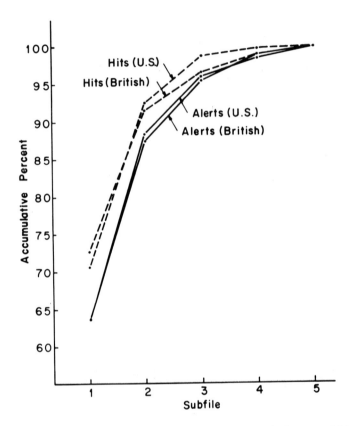

FIG. 10. Comparison of accumulative percents of alerts and hits per subfile for Organic profiles of U. S. and British users.

This looks as if it is contradictory to the results of the average precisions of the ranked subfiles which decrease markedly from one subfile to another. However, it must be remembered that, on the average, the number of citations retrieved by the lower-ranking subfiles is comparatively much lower than those retrieved by the higher-ranking subfiles. So, when each of the values of alerts and hits is added to the previous values, this does not affect much the ratio of total hits to total alerts, even if the individual ratio of hits to alerts is much lower than the previous one.

Results of the comparison of the average and accumulative precisions per subfile for the U. S. and British samples are given in Tables 18 and 19. Graphical representation of these data is given in Fig. 13.

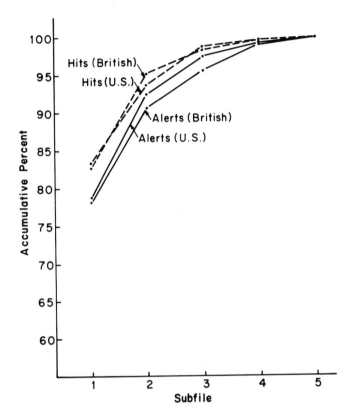

FIG. 11. Comparison of accumulative percents of alerts and hits per subfile for Macromolecular profiles of U.S. and British users.

The results in Tables 13 and 14 (Fig. 16) show that the average preci-sion and accumulative precision per subfile for the British sample is much higher than those of the U.S. sample. The reader is reminded that for each British user, a preliminary search was conducted against a small portion of the file, and the user then adjusted his profile after evaluating the results of the preliminary search. This test was done to explore a better way for the users to interact with the data base. The results in-dicate that, given the opportunity to adjust their profiles, the British users achieved a much higher precision than the U.S. users. However, a final evaluation of this means of user-system interaction would not be possible until the effect of the change in precision on the recall value of the system is analyzed. It is well known that precision and recall generally have an inverse relationship.

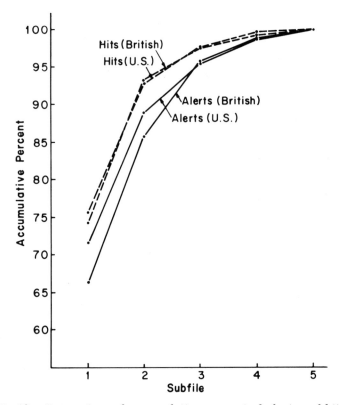

FIG. 12. Comparison of accumulative percent of alerts and hits per subfile for Physical profiles of U.S. and British users.

3. User's Ability to Predict Relevant Subfiles

Having concluded that the material on the topic of interest is mainly localized in two subfiles out of five, it remains important to determine if the user of the file can predict in advance the subfiles that provide him with the most relevant references.

When the users' selections (predictions) of pertinent subfiles were matched against the actual pertinent subfiles, we obtained the results shown in Table 20 for the U.S. and British samples. (See pages 179-182 for the details of the procedure followed in computing the results.)

TABLE 14

Average Precision per Subfile for U.S. Sample

Subfile	1	2	3	4	5
Precision (96 Profiles)	39.9	33.6	19.9	11.4	4.5

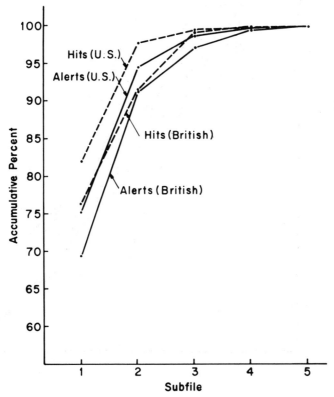

FIG. 13. Comparison of accumulative percents of alerts and hits per subfile for Applied profiles of U.S. and British users.

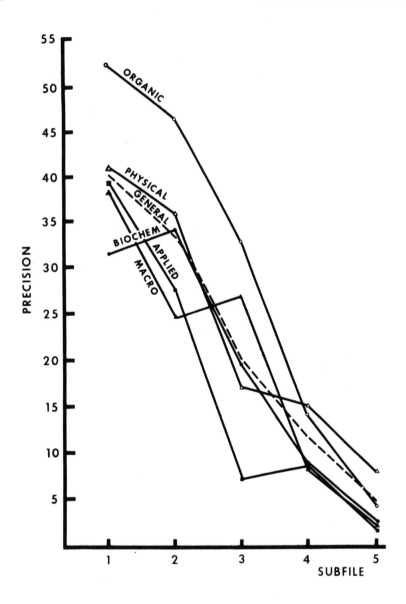

FIG. 14. Average precision per subfile of U.S. sample.

TABLE 15

Average Precision per Subfile for Subject Groups

Biochemistry (19 Profiles)	31.6	34.3	19.6	8.7	2.5
Organic (12 Profiles)	52.2	46.3	32.7	13.7	4.2
Macromolecular (19 Profiles)	38.6	24.5	26.7	7.9	2.1
Applied & Chemical Engineering (12 Profiles)	39.6	27.5	6.9	8.6	1.4
Physical & Analytical (34 Profiles)	40.9	36.0	16.7	14.8	7.9

TABLE 16

Average Accumulative Precision per Subfile for U.S. Sample

Section	1	2	3	4	5
Accumulative Percent (96 Profiles)	39.9	39.0	38.6	37.9	37.7

The following observations can be made on the results in Table 20:

1. Users of the file can predict with very high accuracy the most pertinent subfile in the data base which provides them with the greatest percent of output that is related to their questions (97.9% accuracy for the U.S. sample and 100% accuracy for the British sample).

2. Users' predictions for the second most pertinent subfile are less accurate than the first. However, in many of the cases where the user's predictions for the second pertinent subfile mismatched the actual one, the user selected another subfile (the third most pertinent) with a relatively little less pertinency. That is, in such cases, the number of relevant citations retrieved by the selected subfile (the third most pertinent) was little less than those retrieved by the second pertinent subfile.

3. The same observation as in 2 can be said on the third and fourth pertinent subfiles related to the U.S. sample.

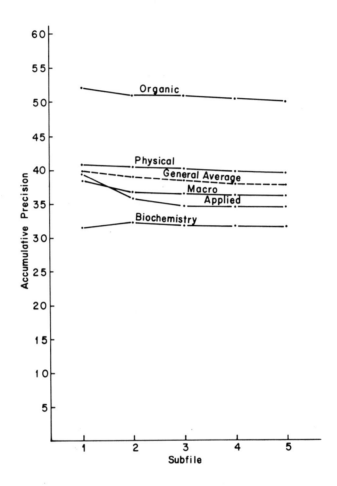

FIG. 15. Average accumulative precision per subfile of U.S. sample.

On the whole, the results showed that the overall weighted average of matches between users' predictions and actual pertinent subfiles are 92.2% for the U.S. sample and 92.9% for the British sample.

Figures 17 and 18 show the respective results of alerts and hits retrieved by the U.S. users' requests to search specific subfiles, and the total percents of alerts and hits (for the ranked subfiles) retrieved by the whole data base. Figures 19 and 20 show the same results for the British users.

TABLE 17

Average Accumulative Precision per Subfile for Subject Groups

Subject Group / Section	1	2	3	4	5
Biochemistry (19 Profiles)	31.6	32.3	31.9	31.8	31.7
Organic (12 Profiles)	52.2	51.1	51.0	50.4	50.0
Macromolecular (19 Profiles)	38.6	36.8	36.5	36.4	36.1
Applied & Chem. Eng. (12 Profiles)	39.6	35.9	34.7	34.6	34.5
Physical & Analytical (34 Profiles)	40.9	40.5	40.2	38.8	38.6

TABLE 18

Comparison of Average Precision per Subfile of
U. S. and British Samples

Subfile	1	2	3	4	5
U.S. Sample (96 Profiles)	39.9	33.6	19.9	11.4	4.5
British Sample (48 Profiles)	57.7	44.0	25.6	14.8	5.6

The following remarks can be said on the results of Tables 14 to 17:

1. The U.S. users' requests to search specific subfiles retrieved 1.0% less alerts and 0.2% less hits than the total number of alerts and hits in the most pertinent subfile of the data base. British users' requests on the other hand, retrieved all the alerts and hits in that subfile.

2. Results of the comparison pertaining to the second, third, fourth, and fifth subfiles indicate that the users tend to request the searching of a less number of subfiles than those which contain material that are related

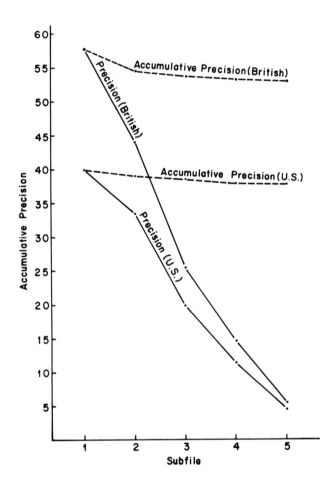

FIG. 16. Comparison of precision and accumulative precision per subfile of U.S. and British profiles.

to their inquiries. As a result, they obtain a smaller percentage of relevant material than is actually present in the file. This material is mainly localized in the least pertinent subfiles of the data base.

The tendency by the users to search a limited number of subfiles is evident from the fact that out of 144 users in the two samples, 66 requested to search only one subfile; 50 requested to search two subfiles; 19 requested to search three subfiles; three requested to search four subfiles; and five requested to search all five subfiles. This in spite of the fact that the service was offered free of charge to all users.

TABLE 19

Comparison of Average Accumulative Precision per
Subfile of U. S. and British Samples

Subfile	1	2	3	4	5
U. S. Sample (96 Profiles)	39.9	39.0	38.6	37.9	37.7
British Sample (48 Profiles)	57.7	54.6	53.8	53.2	53.0

4. Relationship Between the Subjects of Subfiles Accounting for the Largest Percentage of Search Retrievals and Present Arrangement of Subject Groupings in CA Condensates Current-Awareness Tapes

Results of the file subdivision indicated that, on the average, a large percent of the material on topics of interest (91% of alerts and 94% of hits) is localized in two subfiles (out of five) in the data base. A breakdown of the U. S. and British samples of search inquiries (144 profiles) according to the subjects of the two subfiles that retrieved the highest number of alerts and hits (in which the material on their topics of interest is localized) gives the results shown in Table 21.

A comparison was made between the subject combinations of the two subfiles in which the material on topics of interest is localized (Table 21) and the subject combinations of the odd- and even-numbered issues of the Condensates Current-Awareness tapes (which correspond to the Chemical Abstracts issues).* These are:

Issue	Subject Combination
Odd-Numbered	Biochemistry/Organic Chemistry
Even-Numbered	Macromolecular/Applied & Chemical Engineering/Physical & Analytical

The result of the comparison showed that of the combinations of two subfiles which retrieved the largest percentage of alerts and hits (Table 21), only the following subject pairs have both subfiles in the same (odd

---

*The contents of CA Condensates current issues are divided into five section groupings on the basis of which we have divided our retrospective data base. These are published in two consecutive weekly issues: an odd-numbered issue, and an even-numbered issue.

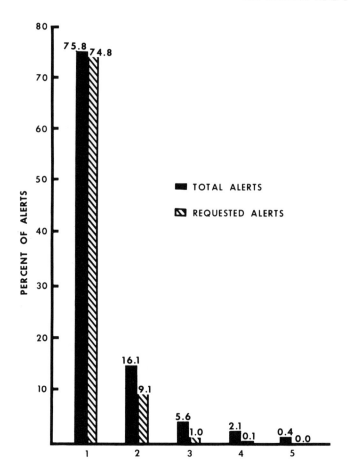

FIG. 17. Comparison of percent of alerts retrieved by users' requests and percent of total alerts in subfiles (U. S. sample).

or even) issues: 1) App/Phys (Applied/Physical), 3) Bio/Org (Biochemistry/ Organic), 5) Mac/App (Macromolecular/Applied), and 8) Mac/Phys (Macromolecular/Physical).

Table 19 shows that the subfiles which are constructed from the above four subject combinations retrieved the largest percentage of alerts and hits for 51.4% of the profiles of the sample (74 out of 144).

In addition, the Biochemistry subfile alone retrieved all the alerts and hits for three more profiles; the Physical Chemistry subfile retrieved

TABLE 20

Ability of Users to Predict Most Related Subfiles

Subfile Rank	First	Second	Third	Fourth	Fifth
Number of Requests	96	50	16	3	1
Number of Matches	94	44	12	2	1
Percent of Matches	97.9	88	75	66.7	100

British Sample

Subfile Rank	First	Second	Third	Fourth	Fifth
Number of Requests	48	28	12	6	5
Number of Matches	48	21	12	6	5
Percent of Matches	100	75	100	100	100

all the alerts and hits for one other profile; and the Macromolecular subfile retrieved all the alerts and hits for another profile.

The above results indicate that if we have decided to build our retrospective data base from the current-awareness tapes so that it was subdivided on the basis of the odd and even issues, only 54.9% of our users would have obtained their alerts primarily from either one of the two subdivisions. The remaining 45.1% would have had their profiles searched against the whole file to retrieve their alerts which are mainly localized in two subgroupings. This supports our conviction that the subdivision of the Condensates file into the five main subgroupings of Chemical Abstracts is a convenient and economical way of retrospective searching of the file.

By the same token, if our sample users were interested in keeping up-to-date on current progress in their areas of interest, 40% of their profiles would have to be searched against both the odd and even issues of the current-awareness tapes. This suggests that it would be more economical for the users if the current-awareness tapes were published in five individual subfiles corresponding to the five main subgroupings.

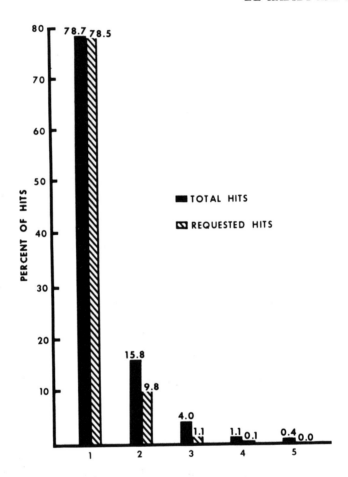

FIG. 18. Comparison of percent of hits retrieved by users' requests and percent of total alerts in subfiles (U.S. sample).

However, this economy might be offset by other factors such as convenience of processing the five separate tapes, and the benefits of batching large numbers of search inquiries.

C. Search System

Computer timings for several searches of profiles against the subfiles of Biochemistry, Organic Chemistry, Macromolecular Chemistry, Applied & Chemical Engineering, and Physical & Analytical Chemistry are given in Tables 22, 23, 24, 25, and 26, respectively. The tables show

TABLE 21

Subject Combinations of Subfiles Retrieving the Highest Percentages of Alerts & Hits

Subject Combination	(1) App/Phys	(2) Org/Phys	(3) Bio/Org	(4) Org/Mac	(5) Mac/App	
Number of Profiles	33	24	23	15	14	Bio 3
Percent	22.9	16.7	16.0	10.4	9.7	Phys 1

Subject Combination	(6) Bio/Phys	(7) Bio/App	(8) Mac/Phys	(9) Bio/Mac	(10) Org/App	
Number of Profiles	11	10	4	4	1	Mac 1
Percent	7.6	6.9	2.8	2.8	0.7	Total 144

CA Odd Issues: (1) App/Phys   (2) Org/Phys

   Bio/Org   (3) Bio/Org   (4) Org/Mac   79 Profiles   54.9%

CA Even Issues: (5) Mac/App   (6) Bio/Phys   65 Profiles   45.1%

   Mac/App/Phys   (8) Mac/Phys   (7) Bio/App

     + Bio, Phys   (9) Bio/Mac

       Mac   (10) Org/App

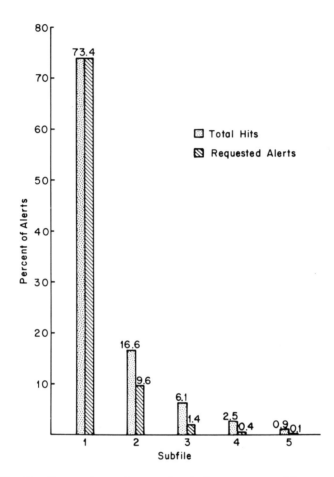

FIG. 19. Comparison of percents of alerts retrieved by users'
requests and percents of alerts in subfiles (British sample).

the relationships between the computer timings and the batch sizes of different queries, number of terms associated with them, size of the subfile
searched, and number of alerts retrieved.

Computer timings given in the above mentioned tables were derived
using an IBM 360/50 computer with 256k of core running under OS with
MVT and HASP II system. In analyzing the data in the above tables, it
was rather difficult to derive some algorithm for the computer timing associated with searching the file, using the TEXT-PAC system, due to the
large number of variables that affect the search system. Examples of

TABLE 22

Retrospective Search Statistics for the Biochemistry Subfile

Section & volume	Number of citations searched	Number of profiles	Total terms	Terms/ profile	Total hits	Hits/ profile	Execution time (min)	CPUs (min)	Computer usage units[a]
Bio 69	28,171	1	15	15	12	12	13.7	3.8	.0933
Bio 69	28,171	10	352	35	55	5	29.1	20.5	.2123
Bio 69	28,171	5	128	35	699	133	25.3	14.6	.2735
Bio 70	34,759	1	15	15	15	15	16.6	4.7	.1129
Bio 70	34,759	3	62	20	146	49	24.9	12.8	.2594
Bio 70	34,759	8	335	41	210	26	45.8	34.1	.4993
Bio 70	34,759	10	246	24	108	10	34.7	22.6	.2529
Bio 71	37,930	1	15	15	12	12	17.7	4.8	.1212
Bio 71	37,930	18	366	20	150	8	46.1	33.1	.3417
Bio 71	37,930	13	381	29	1341	103	44.3	28.2	.3343
Bio 72	41,106	1	15	15	11	11	19.5	5.4	.1329
Bio 72	41,106	13	381	29	1398	105	48.2	31.0	.3628
Bio 72	41,106	3	22	7	676	228	23.0	6.3	.2292
Bio 73	42,706	1	15	15	16	16	21.9	5.7	.1498
Bio 73	42,706	5	201	40	171	33	43.4	29.3	.4668
Bio 73	42,706	3	188	62	656	218	34.5	18.3	.3599
Bio 73	42,706	8	188	23	970	125	44.3	28.1	.3288

[a]Units = (CPU Time + 0.85 IO Time) · (HS Core + 1/2 LSC) / 128 k.

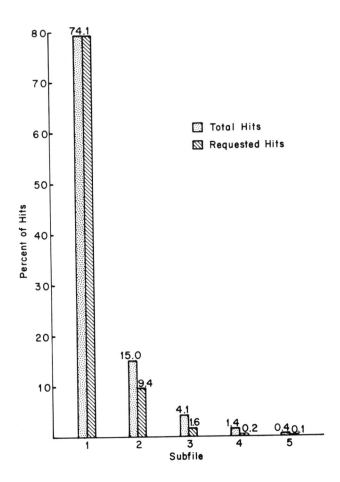

FIG. 20. Comparison of percents of hits retrieved by users' requests and percents of hits in subfiles (British sample).

these variables are: the search logic (or, and, with, adjacent, etc.), the number of search terms, the number of profiles per batch, the number of citations searched, and the number of alerts retrieved. However, an attempt was made to compute the average computer costs associated with searching each of the subfiles in the data base. We averaged the cost of computer search per profile for a sample of 420 computer runs, each consisting of a batch of profiles ranging from 3 to 20. The runs were searched against the different subject subfiles and the different volumes (69 to 73) comprising the data base. The resulting average search costs per profile

TABLE 23

Retrospective Search Statistics for the Organic Subfile

Section & volume	Number of citations searched	Number of profiles	Total terms	Terms/ profile	Total hits	Hits/ profile	Execution time (min)	CPUs (min)	Computer usage units[a]
Org 69	16,659	12	279	23	331	28	17.1	10.0	.1236
Org 69	16,659	12	237	19	268	22	20.6	13.6	.1498
Org 69	16,659	5	94	18	280	56	15.9	8.6	.1124
Org 70	17,850	15	359	23	194	13	23.4	16.1	.1712
Org 70	17,850	15	376	25	343	23	21.2	13.5	.1539
Org 70	17,850	4	84	21	13	3	15.0	8.2	.1062
Org 70	17,850	7	366	52	302	43	32.9	26.0	.2451
Org 71	16,951	2	43	21	22	11	9.7	3.0	.0955
Org 71	16,951	13	269	20	239	18	22.6	15.3	.1652
Org 71	16,951	5	439	87	115	23	18.9	11.9	.1431
Org 72	17,952	13	221	17	291	23	20.8	14.3	.1510
Org 72	17,952	2	43	21	37	18	10.4	3.1	.1024
Org 72	17,952	14	518	36	342	24	43.1	35.9	.3207
Org 73	16,412	3	29	10	44	14	9.5	2.7	.0649
Org 73	16,412	8	334	42	129	18	25.8	19.0	.1903
Org 73	16,412	13	221	17	268	20	19.4	12.6	.1407
Org 73	16,412	21	451	21	511	24	25.3	16.9	.1936

[a]Units = (CPU Time + 0.85 IO Time) • (HS Core + 1/2 LSC) / 128 k.

TABLE 24

Retrospective Search Statistics for the Macromolecular Subfile

Section & volume	Number of citations searched	Number of profiles	Total terms	Terms/ profile	Total hits	Hits/ profile	Execution time (min)	CPUs (min)	Computer usage units[a]
Mac 69	10,401	8	421	52	376	47	19.3	13.2	.1400
Mac 69	10,401	3	89	29	3	1	8.8	4.3	.0618
Mac 69	10,401	10	349	35	85	10	12.9	7.8	.1331
Mac 70	11,812	8	421	52	405	50	21.7	15.2	.1585
Mac 70	11,812	4	158	39	506	126	15.7	9.1	.1123
Mac 70	11,812	10	348	34	105	10	14.4	8.9	.1498
Mac 71	14,430	8	421	52	535	67	27.2	19.7	.1985
Mac 71	14,430	6	461	76	48	8	16.7	10.3	.1201
Mac 71	14,430	10	349	35	114	11	17.3	10.9	.1819
Mac 72	14,513	4	105	26	73	18	14.5	8.4	.1022
Mac 72	14,513	8	421	52	595	74	27.2	19.0	.1983
Mac 72	14,513	5	168	33	583	115	19.3	11.3	.1380
Mac 72	14,513	10	348	34	116	11	17.5	11.0	.1843
Mac 73	14,270	3	158	52	2	--	14.9	8.8	.1057
Mac 73	14,270	3	82	27	75	25	14.3	7.7	.1012
Mac 73	14,270	10	349	34	157	15	17.8	11.0	.1870
Mac 73	14,270	8	421	52	597	74	28.1	19.8	.2051

[a]Units = (CPU Time + 0.85 IO Time) • (HS Core + 1/2 LSC) / 128 k.

TABLE 25

Retrospective Search Statistics for the Applied and Chemical Engineering Subfile

Section & volume	Number of citations searched	Number of profiles	Total terms	Terms/ profile	Total hits	Hits/ profile	Execution time (min)	CPUs (min)	Computer usage units[a]
App 69	18,706	17	323	20	250	16	15.6	23.6	.3320
App 69	18,706	1	67	67	11	11	13.3	5.6	.0924
App 69	18,706	6	279	46	295	49	28.2	20.4	.3047
App 70	21,394	25	490	19	309	12	35.0	26.3	.2642
App 70	21,394	19	414	21	372	19	27.2	19.0	.1986
App 70	21,394	12	356	26	114	9	29.3	21.2	.2147
App 71	24,087	1	67	67	11	11	16.0	7.3	.1116
App 71	24,087	4	84	21	10	2	19.8	11.0	.1406
App 71	24,087	19	414	22	422	22	31.0	21.9	.2264
App 71	24,087	6	354	59	448	73	36.9	26.9	.3998
App 72	25,253	1	67	67	25	25	16.8	7.3	.1168
App 72	25,253	9	270	30	160	17	35.0	25.7	.2570
App 72	25,253	6	354	59	402	67	39.1	28.9	.4246
App 73	23,953	20	421	21	269	13	32.0	10.2	.2401
App 73	23,953	1	67	67	13	13	16.4	7.3	.1145
App 73	23,953	6	354	59	421	70	38.3	29.6	.4156
App 73	23,953	13	390	30	129	9	34.1	24.7	.2500

[a] Units = (CPU Time + 0.85 IO Time) · (HS Core + 1/2 LSC) / 128 k.

TABLE 26

Retrospective Search Statistics for the Physical and Analytical Chemistry Subfile

Section & volume	Number of citations searched	Number of profiles	Total terms	Terms/ profile	Total hits	Hits/ profile	Execution time (min)	CPUs (min)	Computer usage units[a]
Phy 69	39,048	13	317	24	330	25	37.3	24.0	.2732
Phy 69	39,048	27	571	21	530	20	74.7	62.7	.5650
Phy 69	39,048	9	154	17	474	53	32.2	18.8	.2308
Phy 69	39,048	6	304	50	557	92	59.0	44.7	.4402
Phy 70	35,979	13	317	24	298	23	35.0	22.3	.2568
Phy 70	35,979	5	364	72	888	177	36.5	21.8	.2696
Phy 70	35,979	6	304	50	536	89	55.7	41.7	.4143
Phy 71	37,775	2	38	19	6	3	11.5	3.6	.0797
Phy 71	37,775	2	60	30	11	5	26.8	14.4	.1900
Phy 71	37,775	2	119	59	204	102	34.7	21.8	.3702
Phy 71	37,775	6	304	50	487	81	58.2	44.5	.4332
Phy 72	40,515	5	133	27	203	40	29.2	15.1	.2088
Phy 72	40,515	4	76	19	27	7	29.1	15.8	.2078
Phy 72	40,515	2	119	59	209	104	38.5	26.2	.4132
Phy 73	39,687	5	230	46	50	10	38.8	25.5	.2811
Phy 73	39,687	1	136	136	391	391	39.7	24.8	.2889
Phy 73	39,687	2	119	59	216	108	37.5	22.8	.4068

[a]Units = (CPU Time + 0.85 IO Time) • (HS Core + 1/2 LSC) / 128 k.

per volume for the different subfiles are given in Table 27. Costs are based on $150 per computer unit usage.*

Table 27 indicates that the average computer search cost per profile increases linearly with the increase in the size of the data base searches.

D. User Reaction

The data from the follow-up telephone interviews can be used to evaluate the degree to which the retrospective system satisfied users' information needs. The users in the U.S. and Great Britain were interviewed after they received their output and had an opportunity to evaluate it. The results are clearly favorable to the retrospective system. Forty-six interviews were completed for the U.S. group and 34 interviews were completed in Great Britain for the British group. The interview used for the British group was slightly modified to account for their different circumstances.

It was found that 21 of the scientists in the U.S. group became aware of the existence of the retrospective service through an information officer (e.g., librarian) or functionary of one type or another. Fifteen U.S. users reported that they became aware of the availability of the service through a colleague. The remaining ten individuals were alerted to the

TABLE 27

Average Cost of Computer Search per Profile per Volume

Subfile	Size (Percent of citations in subfile to total citations in file)	Average search cost per profile ($)
Biochemistry	28.7	8.11
Organic	13.4	4.28
Macromolecular	10.2	3.39
Applied & Chemical Engineering	17.7	5.21
Physical & Analytical	30.0	8.56

*Computer unit = (CPU Time + 0.8510 IO Time) · (HS Core + 1/2 LSC)/128k.

existence and availability of the system by a memorandum (announcement) circulated throughout relevant areas of the University and at workshop-seminars. All the British users were informed by the United Kingdom Chemical Information Service.

Only one of the 46 U.S. users had no hits, but he was satisfied that the system had done its best and there was no relevant literature covering his narrowly defined area. The vast majority of the remaining 45 U.S. users had very favorable impressions of the service. The few dissenters were unhappy because the literature coverage only went back through 1968, because their broad profiles brought out a lot of noise, or because they were already aware of the existence of most of the articles to which they had been alerted. The favorable reactions ranged through "very satisfied," "useful," "of great benefit," "quite good," and "excellent."

Four of the 34 British users expressed negative opinions about the output they had received. One felt that the precision was too low, another complained about the "messy format," another felt that his search was too broad and gave him too many irrelevant alerts. One indicated that he had a "low opinion" of computerized services in general. The 30 remaining British users expressed positive opinions such as "precise," "excellent," "first class," "worthwhile."

Twenty-nine U.S. users and 16 of the British users had retrospective searches in subject areas related or identical to their current-awareness searches. For both groups, the retrospective search was being used as background for the current-awareness search, or the retrospective search covered an informational subset of the current-awareness search. In the latter case, this sometimes reflected a branching-out of professional interests. Ten U.S. users and 12 British users had search profiles in subject areas that were not related to their current-awareness searches. Most of these were scientists who were starting into new research areas. The remaining seven U.S. users and six British users did not have current-awareness profiles.

Opinion was broadly divided in both groups as to the stage of the research process where the retrospective output would have its greatest functional usefulness. Eighteen U.S. users and only five of the British users felt that it was most useful during the early conceptual stage of research when the scientist was surveying the literature and developing his hypotheses. Thirteen U.S. respondents and 21 Britishers felt that their output was most useful for developing their research design and performing laboratory experiements. Fifteen Americans and five British scientists felt that their hits were most useful for their scientific writing when they could easily make reference to the past work to which they had been alerted. The remaining three British scientists could not decide between any of the stages given, and felt that their output was equally relevant for all stages.

We can see some significant differences between the two groups from these results. Many more U.S. users feel that their output is most useful at either the early conceptual stage of research or at its conclusion when they write their reports. The majority of British users felt that the retrospective service was most useful for the development of their research designs and for guiding their experimentation.

The concept of "relevance" is probably a different analytical dimension with regard to retrospective as opposed to current-awareness literature searching. This is still an intuitive feeling; but the scientist is usually looking for a different, though similar, information base in his retrospective search than in his current-awareness searches. Current-awareness information must be timely and its timeliness will affect its relevance, but the very fact that retrospective information is dated affects its relevance. The relevance estimates from the U.S. and British retrospective users averaged 40.4% and 40.2%, respectively, and these are very close to the relevance figures collected during our earlier current-awareness experiments, i.e., 40% to 44%, generally. The interesting development is that the estimates from the retrospective users show a great deal more variability (0% to 100%) than the estimates from the current-awareness users. Since the sample sizes of the retrospective user groups are still small, it is not really reliable to compute variance statistics for comparison to our much larger current-awareness user group. These data, though weak, still suggest that the large user variability which has already been well documented in the current-awareness experiments may be even larger among the retrospective users.

The information system delivers citations, but the use of the citation does not end there. In terms of information flow concepts, the time when the citation is delivered is really just its beginning for the user. One of the important questions of system evaluation is what becomes of the alert once the user receives it; or in other words, how does the user follow up on relevant citations. Fifteen U.S. users and 15 British users said that they generally consult Chemical Abstracts as a first step, and then go on to read the full-text article. Chemical Abstracts seems to be particularly useful for foreign language publications and for articles which are in more obscure, difficult to obtain, journals. The remaining 31 U.S. users and 19 British users all indicated that they went directly to the full-text articles when they could find them. They did not use Chemical Abstracts as an intermediary step in the information procurement process. For many of the relevant articles followed up in this manner, users either made photocopies of parts or all of each document, or wrote to the author for reprints.

A system should attempt to be as comprehensive as is necessary, given the constraints of its resources. Even when this is accomplished, automated information systems often must be supplemented by manual

literature searching. Thirty of the U.S. users and 19 British users found
this to be needed. They all had to supplement the CASCON retrospective
search for one reason or another. Some felt that the system's output was
incomplete, and that they should check it with a manual search over the
same literature. Other users reported that they supplemented their
search output by going to information sources outside of the standard
chemical literature, e.g., Biological Abstracts, Index Medicus, etc.
Most users who supplemented their output by a manual literature search
did so because the information base of the retrospective system only dates
back to July, 1968, and they needed a search through older documents. Of
course this segment of the user group will continually decrease in number
as the years pass and as more and more dated literature is covered by the
system. Most users of current-awareness services do not depend on them
entirely for being alerted to current literature. It is not surprising, then,
that most users of retrospective services do not put complete reliance on
machine-generated bibliographies.

The question of "how far back" is a serious one to consider in evaluating the effectiveness of a retrospective system. The interviewees were
asked to indicate how far back in the literature they had to search to satisfy their retrospective information needs for the research problem covered
by their computer search. The reader will recall that a similar question
was asked on the initial questionnaire for the retrospective search experiment (Chapter 3). The results of the follow-up interview, thus far, seem
to support the results of the questionnaire. Forty-eight percent of the
U.S. interviewees and 32% of the British interviewees only had to search
back within a five-year period. On the other hand, 37% of the U.S. users
and 59% of the British users indicated that they had to search back over
nine or more years of the chemical literature. The British group indicated an interest in more dated literature.

Some general remarks about the overall popularity of the retrospective literature searching service are in order at this point. Forty-one of
the 46 U.S. users reported that they had recommended the use of the service to at least one of their colleagues while only 12 of the 34 British users
had done so. Thirty-eight of the 46 U.S. users and 27 of the 34 British
users indicated that they would need and use the system in the future. This
is an important factor to consider in the evaluation of a retrospective
search system. Since a particular search is a one-time service and a
one-time cost, it seems as if the return rate of users to the system a
second, third, or fourth time would be a good indicator of user satisfaction and system performance. As a scientist's interests change and as
his research problems either branch out or change entirely, he should
be drawn back to the retrospective system if his first experience was a
rewarding one.

Of course, it would be misleading to say or to imply that every user was totally happy and satisfied with the service. Some users did report problems, but they were very few in number. One user reported that the profile was too broad and very difficult to narrow down to more manageable dimensions. As a result, he received many irrelevant alerts and found this disturbing. One user complained about the nomenclature problem he encountered in constructing his search strategy. Some of the members of the British group complained about not having left-truncation facilities. One user even complained that the address of the author should be included on the alert so that he could write for reprints more easily.

The telephone interview has proven to be a very effective method of data collection when the interviewer's purpose is to obtain answers to some well-defined questions. If the respondent has the information in hand to answer the questions, and the time required to answer the full set of questions is kept to a minimum (our interviews generally lasted about five minutes), the telephone interview is probably the best method of data collection used on the project. It provided a large amount of good data in a relatively short period of time, with a very high response rate, and minimum of effort on the part of the interviewer and interviewee.

1. User Reactions to Costs

We have two sets of data about the costs of the retrospective literature searching service. The service was offered on a non-fee paying (free) basis. Therefore we had to rely upon the subjective estimates of the users to determine the value of the service in terms of dollar amounts. The first set of data deals with the users' expectations about the costs of the service. These data were derived from the responses to a question asked on the initial questionnaire that U.S. and British users completed before submitting their search strategies. Users were therefore responding according to their preconceived notions about the system's worth.

We expected that the amount of money that a scientist would be willing to allocate from his research budget to pay for the service would be affected by his total budget. It was evident that we should control for this effect. Each respondent was asked to first indicate the size of his operating research budget (to the closest $5000) and then asked how much he would be willing to allocate from his budget for the service. In this way, we could treat the users' cost estimates as percentages of their total research budgets.

Before reporting these data (Table 28), we wish to make clear that they are not as strong as we would have preferred. For several reasons, many of the questionnaire respondents did not answer this question. Only 28% (46 out of 164) of the U.S. respondents and 16% (18 out 114) of the

TABLE 28

Percent of Research Budget the User is Willing to
Allocate for Retrospective Searches

	Number responding	
Percent	U. S.	G. B.
0	5	1
less than 1	11	7
1	10	4
2	8	1
3	7	2
4	5	0
5	0	3
	46	18

British respondents answered the question satisfactorily. There were two
major reasons for not responding. Many of the U.S. users were graduate
students and had no direct knowledge of research resources. Secondly,
many users in both the U.S. and Great Britain worked for industrial firms
in which it was felt that the size of their scientists' research budgets was
proprietary information and could not be indicated in response to this
question. The overall reason for both groups was a general confusion about
the issue of costs.

The charges for the service should obviously take into consideration the
users' ability to pay. If it is decided that $100 per search will be charged
for the service, the salesman should start looking for scientists with research budgets that can support such fees.

In addition to these figures reflecting expectations, the users were
asked to respond to an interview question concerning the costs after they
had received and had a chance to evaluate their output. To this effect, they
were asked to report the value, in dollars, of their output; in other words,
how much they would be willing to pay for what they received. Of the 46
U.S. interviewees who received hits, ten felt that they could not estimate
the value of their output and could not answer the question. The remaining

36 were fairly close in their estimates of value. The average of their estimates was $97 and the range from $25 to $250. Of the 34 British interviewees, ten felt that they could not estimate the value of their output and could not answer the question. The remaining 24 were again fairly close in the estimates of value. The average of their estimates was £50 ($125) and the range was from £10 to £150 ($25 to $375).

Obtaining sound data on the costs of services such as these is extremely difficult for several reasons. The whole phenomenon of paying relatively large sums of money for scientific information is a very new development in the recent history of science. Somehow libraries have always been "free" (or at least appeared to be so) and automated information retrieval was thought by some to be an extension of this library service. Secondly, it is extremely difficult to quantify (in dollars, particularly) a new information source, especially when it provides a service that the user had previously been able to supply for himself. The most important value of computer-based information is that it saves time and it is difficult at first to assess how large this saving is overall. Finally, many information services have found it difficult to prove their worth in money while still in the design and development stage when system reliability is not always good.

V.   CONCLUSIONS AND RECOMMENDATIONS

The most important goal of the PCIC retrospective search experiment, discussed in this report, was to develop an economic means for the retrospective searching of CA Condensates based on a system designed to meet the special needs and requirements of its users. For this purpose, the focus of the study was on two main elements: 1) to determine the basic characteristics of information requirements for the retrospective user group by characterizing the expectations which scientists have of the system, and then to evaluate the degree to which the system has satisfied these needs; and 2) to design a system capable of retrospective searching a large data base, such as CA Condensates, economically and effectively.

Another significant result of the experiment is its contribution to international research in chemical information retrieval since the participation in the experiment was extended to some of the users in Great Britain through the United Kingdom Chemical Information Service (UKCIS). Both the British group and the UKCIS staff were most cooperative throughout the experiment. Such successful cooperation leads us to recommend the extension and continuation of such efforts to promote the development of improved systems for handling computer-based information. International cooperation in the field of mechanized chemical information is necessary to give the broadest benefit from the modern technology that has been brought to bear in the retrieval of chemical information.

We have noted that the interest of users in retrospective searching was very slow in developing when the service was first offered only to those PCIC users who had participated in the earlier current-awareness service. However, when participation in the experiment was extended to a new sample of users, the service was very well received in both the U. S. and Great Britain, with few or no reservations. This validates our hypothesis that most of the current-awareness users had already satisfied their information needs from the current-awareness tapes, in that most users thought that retrospective searching would provide them with redundant information, especially since the file goes back only to July, 1968.

The fact that 300 users responded to the invitation to participate in the experiment is quite significant. Although it was not possible to derive a true random sample from all the population of chemists, we feel that the sample of users from both the United States and Great Britain was representative of the major sectors of the chemical research communities, i. e. , academic, industry, non-profit research, and government, the possible users of the computerized file in the near future.

In developing a large-scale retrospective search capability there are two main elements which influence the system economics: first, software development requirements; and second, file organization.

The economics of information handling largely depends on the availability of the necessary software, and one of the major problems facing the system designer is whether to develop the required software for himself or to acquire a preprogrammed package. Lack of experience in large database management, particularly in chemical information, at the time of developing our system led us to favor the use of a prewritten program, i. e. , IBM TEXT-PAC system in the early stages of the PCIC experiment. Although TEXT-PAC was not specially designed for searching chemical information, its unique organization through a semi-inverted file makes it suitable for searching large files such as CA Condensates. However, a great deal of work and file manipulation was necessary to implement the search system, because it is very large and intricate, and the documentation was not very well organized since it has been made available in a prerelease version of the program. We are grateful to Dr. S. Kaufman of IBM for his help during implementation and are confident that new users of the TEXT-PAC system will have relatively little difficulty in using it.

A. Economics of File Organization

One of the principal difficulties in developing large-scale retrospective searching is the economics of searching an extremely large file such as CA Condensates. To reduce the cost of searching the file, some criteria for partitioning the data base are needed.

Probably the major conclusion of our experiment is that an economical approach to retrospective searching was achieved through subdividing the data base on a subject basis. Our hypothesis that the material on the topic of interest is localized in a comparatively few subject divisions of the file has been supported by the results of the experiment for the U.S. sample which showed that, on the average, 91.4% of alerts and 94.5% of hits are contained in two subfiles (subject divisions) out of five. When these results were tested on another sample (the British sample), it was found that there is a marked similarity in the results of the two groups (91.4% alerts and 90.5% hits for the U.S. sample against 90.5% alerts and 94.1% hits for the British sample). This further supports our conclusion and indicates that the results are stable and can be applied on a large scale.

Results of precision through the ranked subfiles, which showed that there is a marked difference between the precisions of the first two subfiles (39.9 and 33.6, respectively) and the last three (19.9, 11.4, and 4.5) further supports our conclusion. However, it must be remembered that the results differ from different topics on different subject areas, and this has to be taken into consideration in any application of these findings.

The results of the experiment indicated that users are able to predict in advance the relevant subfiles in which the material related to their inquiries are localized. There were, respectively, 97.9% and 100% matches between U.S. and British users' predictions for the most relevant subfile in the data base and the actual most pertinent subfile. On the whole, the weighted average of matches between users' predictions and actual pertinent subfiles were 92.2% for the U.S. sample and 92.9% for the British sample. These results support our second hypothesis that users of the file have some experience and knowledge of the subject contents of CA groupings which will enable them to predict which subfiles will account for the greatest percent of their output.

However, the results showed that, in selecting the subfiles they would like to search, users tend to request the search of a limited number of subfiles. As a result, they retrieve a smaller percentage of material related to their inquiries than actually present in the file.

In view of the favorable results from our approach to subdividing the file on a subject basis, it is suggested that similar studies on the 80 sections of Chemical Abstracts might be useful.

B. User Needs

Although there were few significant differences between the two groups (U.S. and British), we can feel confident that their reactions to the retrospective service were generally similar.

The evaluative research effort of the behavioral task group was successful on the international level. The data gathering instruments (questionnaires and interviews) were kept simple and the response was very favorable. The British group had a high response rate (90% +) on both instruments and were most cooperative. The staff at Nottingham was able to administer the interview to their users and to elicit good responses with little or no difficulty. This has implications for the future since the science information problem has become world-wide, and more international research efforts and programs will have to be established to provide solutions which will work in many or all of the technological centers of the world.

The most important characteristic of user information requirements is that many scientists have retrospective information needs that go back further than the information file of the system (July 1968). This too has important consequences for the future, because as time passes and more literature is included in the file, its value as a service will increase for the research scientist. The retrospective system will be able to successfully serve a larger proportion of the scientific community. It was found that the retrospective users were mostly using the system "to complete the coverage of a research area with which they were already very familiar" and many of them were submitting retrospective searches in areas similar or identical to their current-awareness searches. It is expected that the value of the system will also grow over time as a means of providing a comprehensive background for the scientist's current research interests. Our results also support the contention that libraries are an integral part of any information system. Many users had to supplement their retrospective searches with manual literature searches in their libraries. They were also alerted to a large volume of literature which depended greatly for its retrieval on efficient library procedures.

We have found that the availability of the retrospective service must be well publicized in order to attract users. The scientific information grapevine was not sufficient. Most users became aware of the availability of the service from the librarians or information officers in their organizations or by the publicity and workshop seminars that were sponsored by the Pittsburgh Chemical Information Center. Users must be informed and reinformed about these services in order to elicit their acceptance. The impact of the service on the day-to-day activities of the research scientist has been difficult to detect, but the fact that it has affected their general orientations to the chemical literature has been documented.

ACKNOWLEDGMENT

We are indebted to the staff of the United Kingdom Chemical Information Service at the University of Nottingham for their valuable cooperation in arranging the participation of users from Great Britain in our experiment. We are particularly grateful to Mrs. Francis Barker, Research Officer at the Nottingham Center, for coordinating the experiment at Nottingham and carrying out all liaison with the British users.

REFERENCES

1. Chemical Abstracts Service, "Report on the Thirteenth Chemical Abstracts Service Open Forum, Toronto, Canada, May 24, 1970," CAS, Columbus, 1971.
2. J. K. Park et al., "The Development of a General Model for Estimating Search Time for CA Condensates," J. Chem. Doc., 19(4), 1970, p. 282.
3. F. W. Lancaster, Evaluation of the Medlars Demand Search Service, National Library of Medicine, Washington, D.C., 1968, p. 167.

INDEX

Numbers in parentheses are reference numbers.

A

Administration, Pittsburgh Chemical Information Center, 23-24
Alert response cards, 58-59
Applied and chemical engineering subfile, retrospective search statistics, 199ff
Applied research
  computer-based information service, 4-5
  user study, 70-75
Arnett, E.M., 84(1, 2), 122
ASCA, user study, 66-67
Association of Scientific Information Dissemination Centers, 35
Atherton, P., 126, 137
Audacious, interactive retrieval system, 126

B

Badger, G., 91(6), 122
Basic research
  computer-based information service, 4-5
  user study, 70-75
Behavioral Research Group, 25
Behavioral science, role, 44-45
Bennett, J.L., 127(7), 137
Bergman, S., 135(20), 138
Berul, L.H., 135(21), 138
Biochemistry subfile, retrospective search statistics, 195ff

Bloemeke, M.J., 92(8), 122; 128(10), 138
Browsing, and research style, 7
Budget changes, and attitudes toward chemical information, 13
Bush, V., 135(18), 138

C

Calhoun, J.B., 136, 138
Campus-based information system, interactions, 42
Caponio, J.F., 136(24), 138
Caruso, D.E., 126, 127(8); 129(12); 130(13); 133(15, 16), 137, 138
Chemical Abstracts
  collective indexes, size, 141
  sections, 87-89
Chemical Abstracts Condensates
  construction of subfiles, 144-149
  data base description, 86-91
  interactive search, 129ff
  retrospective searching, 110-116, 139-211
  search package, implementation in PCIC program, 84-123
  user study, 64-67
Chemical Abstracts Service, 10-12
Chemical Information Center, University of Nottingham, interactions, 35-36
Chemical information science, aims, 8

Other books of interest to you...

Because of your interest in our books, we have included the following catalog of books for your convenience.

Any of these books are available on an approval basis. This section has been reprinted in full from our *library science/ information science/computer science* catalog.

If you wish to receive a complete catalog of MDI books, journals and encyclopedias, please write to us and we will be happy to send you one.

MARCEL DEKKER, INC.
95 Madison Avenue, New York, N.Y. 10016

library science
information science
computer science

ARNETT and KENT *Computer Based Chemical Information*

(Books in Library and Information Science Series, Volume 4)

edited by EDWARD ARNETT, *Department of Chemistry, University of Pittsburgh, Pennsylvania,* and ALLEN KENT, *Director, Office of Communication Programs, University of Pittsburgh, Pennsylvania*

280 pages, illustrated. 1973

Concerned with the development of a variety of computer–searchable indexes. Describes experiences arising from a major experimental program in accessing some of these data bases in order to provide them to a large group of chemists, and includes behavioral studies on the use of the literature by chemists before and after the introduction of these services. A great aid to librarians, chemical information officers, research directors, and research chemists.

CONTENTS: The research chemist and his information environment, *E. Arnett and A. Kent.* The Pittsburgh Chemical Information Center—internal and external interactions, *E. Arnett and A. Kent.* The user's information system: An evaluative research approach, *D. Amick.* System design, implementation, and evaluation, *N. Grunstra and J. Johnson.* Interactive retrieval systems, *E. Caruso.* Chemistry library, *M. Roppolo.* Approaches to the economical retrospective machine-searching of the chemical literature, *B. El-Hadidy and D. Amick.*

BEKEY and SCHWARTZ *Hospital Information Systems*

(Biomedical Engineering Series, Volume 1)

edited by GEORGE A. BEKEY, *University of Southern California, Los Angeles,* and MORTON D. SCHWARTZ, *California State College, Long Beach*

416 pages, illustrated. 1972

Brings together the current knowledge on hospital information systems, previously scattered in diverse publications involved in the implementation of hospital information systems. Among the main areas cov-ered are: the clinical laboratory, the hospital ward, the intensive care unit, the coronary care unit, the pharmacy, the multi-test screening center, and the business office. Of vital interest to hospital administrators, biomedical engineers, medical computer research workers, architects, physicians, nurses, paramedical workers, and all others concerned with the modernization of medical care.

CONTENTS: Review of hospital information systems, *E. C. DeLand and B. W. Waxman.* Status of hospital information systems, *M. D. Schwartz.* Electronic data processing of prescriptions, *R. F. Maronde and S. Seibert.* On-line data bank for admissions, laboratories, and clinics, *G. E. Thompson.* Hospital information system effectiveness, *E. J. Bond.* Data processing techniques for multitest screening and hospital facilities, *M. F. Collen.* Automation and computerization of clinical laboratories, *M. D. Schwartz.* Computing systems in hospital laboratories, *M. J. Ball, J. C. Ball, and E. A. Magnier.* Computer applications in acute patient care, *M. D. Schwartz.* A computer-based information system for patient care, *H. R. Warner.* Use of automated techniques in the management of the critically ill, *M. H. Weil, H. Shubin, L. D. Cady, Jr., H. Carrington, N. Palley, and R. Martin.* Retrospect and overview, *G. A. Bekey.*

DAILY *The Anatomy of Censorship*

(Books in Library and Information Science Series, Volume 6)

by JAY E. DAILY, *Graduate School of Library and Information Sciences, University of Pittsburgh, Pennsylvania*

400 pages, illustrated. 1973

An analysis of the motives of the censor, rather than a history of censorship. Exposes the essential purposes of censorship: the maintenance of a propaganda line, the preservation of stereotypes, and the limitation of knowledge.

CONTENTS: Don't touch my dirty words • The horses of instruction • The tigers of wrath • The road of excess • The palace of wisdom • The awakened sex • Sensuous education • To turn a trick • Intellectual freedom and the world community.

DAILY *Organizing Nonprint Materials*

(Books in Library and Information Science Series, Volume 3)

by JAY E. DAILY, *Graduate School of Library and Information Sciences, University of Pittsburgh, Pennsylvania*

200 pages, illustrated. 1972

Presents the means for organizing collections of nonprint material so that the highest level of efficiency combines with the most effective service to the community. Provides a critical introduction to the understanding of both the problems and possibilities of organizing nonprint collections. Examines specialized tools used in organizing picture collections, and provides a list of uniform titles, a list of subject headings for phonorecordings, and other examples of cataloging nonprint material. Also includes an extensive bibliography to guide the reader to supplementary information.

CONTENTS: Nonprint materials, a problem of definition • Defining the library and its patrons • Pictures and other nonprint material that contains no self-description • Recorded sound • Motion pictures • The procedural manual for nonprint materials • Community survey • Community survey of the Media Center of the Graduate School of Library and Information Sciences of the University of Pittsburgh • Procedural manual • Sources for audiovisual materials • Cataloging phonorecordings • Exogenous description • Classified list of subject headings • Alphabetized list of subject headings.

MANHEIMER *Style Manual: A Guide for the preparation of Reports and Dissertations*

(Books in Library and Information Science Series, Volume 5)

by MARTHA L. MANHEIMER, *Graduate School of Library and Information Sciences, University of Pittsburgh, Pennsylvania*

200 pages, illustrated. 1973

Designed to resolve the problems of anyone working towards a degree and also of value to other students preparing dissertations and term papers who have found other style manuals inadequate. Provides models of bibliographic references for all types of material commonly referred to, and establishes patterns for internal format used to structure the content of the dissertation. Highly useful for any scholarly paper in the field of library and information science and related areas.

CONTENTS: Introduction • Quotations • Bibliographic references • Authorship or entry element • Corporate authorship and documents • Periodicals, other serials, and archives • Chapter references and the bibliography.

WILLIAMS, MANHEIMER, and DAILY *Classified Library of Congress Subject Headings*

(Books in Library and Information Science Series, Volumes 1 and 2)

by JAMES WILLIAMS, MARTHA L. MANHEIMER, and JAY E. DAILY, *Graduate School of Library and Information Sciences, University of Pittsburgh, Pennsylvania*

Vol. 1 *Classified List*
296 pages, illustrated. 1972

Vol. 2 *Alphabetic List*
512 pages, illustrated. 1972

Provides, for the first time, a complete and consecutive listing of all Library of Congress suggested classification numbers along with their corresponding subject headings. Volume 1 contains a list of classification numbers in classified order, together with their subject headings (including subheadings and sub-subheadings) that appear in the 7th edition of the Library of Congress List. Volume 2 consists of an alphabetical listing of the classified subject headings contained in Volume 1, with their respective classification numbers. In this volume, unused or obsolete headings are cross referenced with the legal classified subject headings.

The reader can locate in the alphabetical listing the general category of the book to be cataloged, find this category in the classified list, and then easily choose its exact subject heading. Absolutely essential volumes for every cataloger, librarian, and library school. Also of value to researchers in any field who need to relate subject headings directly to classification numbers.

——————— **OTHER BOOKS OF INTEREST** ———————

MATTSON, MARK, and MacDONALD
Computer Fundamentals for Chemists

(Computers in Chemistry and Instrumentation Series, Volume 1)

edited by JAMES S. MATTSON, *Rosenstiel School of Marine and Atmospheric Sciences, University of Miami, Florida,* HARRY B. MARK, JR., *Department of Chemistry, University of Cincinnati, Ohio,* and HUBERT C. MACDONALD, JR., *Koppers Company, Inc., Monroeville, Pennsylvania*

384 pages, illustrated. 1973

MATTSON, MARK, and MacDONALD
Electrochemistry: Calculations, Simulation, and Instrumentation

(Computers in Chemistry and Instrumentation Series, Volume 2)

edited by JAMES S. MATTSON, *Rosenstiel School of Marine and Atmospheric Sciences, University of Miami, Florida,* HARRY B. MARK, JR., *University of Cincinnati, Ohio,* and HUBERT C. MACDONALD, JR., *Koppers Company, Inc., Monroeville, Pennsylvania*

488 pages, illustrated. 1972

MATTSON, MARK, and MacDONALD
Spectroscopy and Kinetics

(Computers in Chemistry and Instrumentation Series, Volume 3)

edited by JAMES S. MATTSON, *Rosenstiel School of Marine and Atmospheric Sciences, University of Miami, Florida,* HARRY B. MARK, JR., *Department of Chemistry, University of Cincinnati, Ohio,* and HUBERT C. MACDONALD, JR., *Koppers Company, Inc., Monroeville, Pennsylvania*

352 pages, illustrated. 1973

SACKS *Measurements and Instrumentation in the Chemical Laboratory*

by RICHARD D. SACKS, *Department of Chemistry, University of Michigan, Ann Arbor*

in preparation. 1973

SACKS and MARK *Simplified Circuit Analysis: Digital Analog Logic*

by RICHARD D. SACKS, *University of Michigan, Ann Arbor, and* HARRY B. MARK, JR., *University of Cincinnati, Ohio*

176 pages, illustrated. 1972

——————— **ENCYCLOPEDIAS OF INTEREST** ———————

ENCYCLOPEDIA OF COMPUTER SCIENCE AND TECHNOLOGY

edited by JACK BELZER, ALBERT G. HOLZMAN and ALLEN KENT, *University of Pittsburgh, Pennsylvania*

Subscription Price: $50.00 per volume

Single Volume Price: $60.00 per volume

in preparation. 1973

This *Encyclopedia* covers every aspect of computer technology and the fields in which computer technology is used. The scope is developed by the enumeration of an alphabetical list of topics ranging from the specific to the generic. It examines the history and development of computers, the current state of computer science, and the role computers will play in our society in the future. Treatment of subject matter is straightforward, yet scholarly and exhaustive, so that the articles are both comprehensible to the layman and stimulating to the informed specialist. The *Encyclopedia* is of utmost interest and utility to computer hardware specialists, programmers, systems analysts, operations researchers, and mathematicians. •

Volume 1: *Abacus to Arithmetic*

CONTENTS

Abacus, *J. Belzer.* Abstract Algebra, *M. Perlman.* Abstracting, *B. Mathis and J. Rush.* Abstracting Services, *M. Weinstock.* Acceleration Methods, *J. Wimp.* Access and Accessing, *P. Meissner and F. Martin.* Accounting for Computer Usage, *E. Miles, Jr.* Ac-

(continued)

counting, Computers in, *D. Li.* Accumulator, *T. Trygar.* Accuracy, *G. Moshos.* Acoustic Memories, *C. Smith.* Adaptive and Learning Systems, *W. Jacobs.* Address—Addressing, *J. Belzer.* Administrative Systems and Data Processing, *D. Teichroew.* Advanced Planning, Computers in, *H. Linstone.* The ARPA Network, *H. Frank.* The Aerospace Corporation Digital Computing Capability, *R. Van Vranken.* Airline Reservation Systems, *T. Wendel.* Alexander, Samuel, *M. Fox.* Algeria, Data Processing in, *Y. Mentalecheta.* ALGOL, *J. Belzer.* Algorithm, *A. Anderson.* Allocation Models, *S. Elmaghraby and M. El-Kammash.* Alphabets, *M. Grems.* Alpha systems, *A. Ershov.* American Chemical Society—Chemical Abstracts Service, *G. Gautney, Jr. and R. Wigington.* American Federation of Information Processing, *B. Gilchrist and R. Tanaka.* American Institute of Certified Public Accountants, *N. Zakin.* American Institute of Civil Engineers with Emphasis on the Use of Computers, *J. Cobb, Jr.* American Institute of Industrial Engineers, Computers in, *J. Bailey and V. Sahney.* American Institute of Physics, *R. Lerner.* American Management Association, *J. Enell and J. Alexander.* American Mathematical Society, *G. Walker.* American Records Management Association, *R. Grimes.* American Society of Civil Engineers, Computers in, *J. Fleming.* American Society for Information Science, *R. McAfee.* American Statistical Association, Computers and Statistics, *A. Goodman.* Amplifiers, Operational *K. Oka.* The AN/FSQ-7, *E. Wenzel.* Analog Signals and Analog Data Processing, *W. Karplus.* Analog-Digital Conversion, *B. Stephenson.* Analysis of Variance, *H. Ginsburg.* "AND" or "NOT", "NAND" and "NOR" logic, *M. Mickle.* Annual Review of Information Science and Technology, *C. Cuadra.* APL Terminal System, *H. Katzan, Jr.* Approximation Methods, *G. Byrne.* APT (Automatically Programmed Tools), *J. Goodrich and R. Thrush.* Argentina, Computers in, *M. Milchberg.* Argonne National Laboratory Computer Center, *M. Butler.* Arithmetic Operations, *C. Donaghey.*

ENCYCLOPEDIA OF LIBRARY AND INFORMATION SCIENCE

in multi-volumes

editors: ALLEN KENT and HAROLD LANCOUR

assistant editor: WILLIAM Z. NASRI
Graduate School of Library and Information Sciences and The Knowledge Availability Systems Center, University of Pittsburgh

Subscription price: $40 per volume

Single volume price: $50 per volume

ADVISORY BOARD

Olga S. Akhamova, *U.S.S.R.* • Jack Belzer, *U.S.A.* • Charles P. Bourne, *U.S.A.* • Douglas Bryant, *U.S.A.* • David H. Clift, *U.S.A.* • Jay E. Daily, *U.S.A.* • Robert B. Downs, *U.S.A.* • Sir Frank Francis, *England* • Emerson Greenaway, *U.S.A.* • Cloyd Dake Gull, *U.S.A.* • Shigenori Hamada, *Japan* • J. Clement Harrison, *U.S.A.* • Donald J. Hillman, *U.S.A.* • J. Phillips Immroth, *U.S.A.* • William V. Jackson, *U.S.A.* • B. S. Kesavan, *India* • Preben Kirkegaard, *Denmark* • W. Kenneth Lowry, *U.S.A.* • Kalu Okorie, *Nigeria* • E. Pietsch, *Germany* • S. R. Ranganathan, *India* • Samuel Rothstein, *Canada* • Nasser Sharify, *U.S.A.* • Marietta Daniels Shepard, *U.S.A.* • Vladimir Slamecka, *U.S.A.* • Mary Elizabeth Stevens, *U.S.A.* • Roy B. Stokes, *England* • C. Walter Stone, *U.S.A.* • Josef Stummvoll, *Austria* • Orrin E. Taulbee, *U.S.A.* • Lawrence S. Thompson, *U.S.A.* • Eileen Thornton, *U.S.A.* • L. J. van der Wolk, *Holland* • B. C. Vickery, *England* • Bill M. Woods, *U.S.A.*

Up to this time there has not been available a single source containing a comprehensive and unified treatment of the fields of both library science and information science. This multi-volume *Encyclopedia* is the first complete and authoritative work on library science and information science, combining both theory and practice of the two fields in the United States and abroad. The *Encyclopedia* provides a source of easy access, ready reference, and comprehensive coverage of the concepts, terms, methods, and important personages of the two fields, and displays and enhances the essential interdependence of the two areas.

Volume 9: Fore-Edge to Grabhorn   700 pages, illustrated. 1973

CONTENTS

Fore-Edge Painting, *A. Skoog* • Forest Press, Inc., *R. Sealock* • Forgeries, Frauds, Etc., *L. Thompson* • Format, Catalog, *J. Daily* • FORTRAN, *J. Williams* • France, Libraries in,

P. Salvan • France, Library and Information Science, Current Issues in, *E. de Grolier* • Frankfurt Book Fair, *S. Taubert* • Franklin, Benjamin, *W. Nasri* • Franklin Book Programs, *J. Daily* • Franklin D. Roosevelt Library, *J. Marshall* • Franklin Institute Library, *E. Hilker* • Free Libraries, *S. Jackson* • Free Library of Philadelphia, *K. Doms* • Friends of Libraries, *S. Wallace* • Fugitive Materials, *N. Bowman* • Funding—Library Endowments in the United States, *W. Jackson* • G.E. 250 Information Searching Selector, *A. Kent* • Game Theory, *A. Holzman* • Garamond, Claude, *J. Dearden* • Genealogical Libraries and Collections, *G. Doane* • General Semantics, *C. Read* • Gennadi, Grigorily Nikolaevich, *O. Akhmanova* • Geographical Codes, *K. Kansky* • Geographical Libraries and Map Collections, *J. Wolter* • Geographical Literature, *N. Corley* • Geological Libraries and Collections, *G. Lea, P. Briers, and A. Harvey* • Geological Literature, *G. Lea, J. Diment, and A. Harvey* • George Peabody College, School of Library Science, *E. Gleaves* • (The George Washington University, the Medical Center), Biological Sciences Communication Project, *R. Wise* • Georgia Institute of Technology, School of Information and Computer Science, *V. Slamecka* • Georgia Library Association, *D. Estes* • German Union Catalog, *R. Stueart* • Germany, Libraries and Information Centers in, *E. Pietsch, H. Fuchs, H. Ernestus, D. Oertel, M. Cremer, H. Arntz, C. Muller, K. Buschbeck, and K. Schubarth-Engelschall* • Ghana, Libraries in, *E. Amedekey* • Gifts and Exchanges, *J. DePew* • Glasgow, University of Glasgow Library, *R. MacKenna* • Golden Cockerel Press, The, *R. Cave* • Gorbunov-Posadov Ivan Ivanovich, *O. Akhmanova* • Government Publications (Documents), *J. Childs* • Government Printing Office, *C. Buckley* • Grabhorn Press, *R. Bell*

More detailed information on our encyclopedias is available upon request.

Marcel Dekker sectional catalogs *are available in the following disciplines:*

analytical chemistry / spectroscopy

biochemistry / biology / biophysics / clinical chemistry

environmental science

inorganic and organic chemistry

library science / information science / computer science

material science

mathematics / statistics

medicine / drugs

physical chemistry / physics / surface and colloid chemistry

A copy of any sectional catalog in the above areas, as well as a complete catalog may be obtained by sending your request to Marcel Dekker, Inc.